Computer Programming for Spatial Problems

E Bruce MacDougall

The University of Toronto

Edward Arnold

First published 1976 by
Edward Arnold (Publishers) Ltd
25 Hill Street, London W1X 8LL

ISBN paper: 0 7131 5865 4

Printed in Great Britain by William Clowes & Sons, Limited, London, Beccles and Colchester

Preface

The premises of this book are that spatial problems tend to have characteristics which require a somewhat different approach to computer programming than conventional scientific problems, and that it is essential for students, researchers, and professionals dealing with such problems to have sufficient knowledge of these particular characteristics to be able to write small programs of their own, modify existing programs written for other purposes, or direct and advise professional programmers in such tasks.

A central and integral part of this book is a set of example computer programs (in FORTRAN) dealing primarily with mapping. These programs are intended to be of practical as well as explanatory value in that they are short enough to be quickly punched on cards for use with real data. The reader is cautioned, however, against regular application of these programs with major data sets without the addition of coding to check the parameters of particular problems to ensure that the necessary assumptions have been met.

This book has been written after ten years of teaching and research on the topic, first as a graduate student and faculty member in geography at the University of Toronto, then as a faculty member in the Graduate School of Design at the University of Pennsylvania, and finally as an Associate Professor jointly appointed in the Faculty of Forestry and Landscape Architecture, and the Department of Geography at the University of Toronto. Many colleagues and students have influenced me during this time, but I would like to give special thanks to students and faculty of the Department of Landscape Architecture and Regional Planning at the University of Pennsylvania. Most of the ideas described in this book are a direct result of my discussions with them. I would like to give particular thanks to Meir Gross, who detected several bugs in the example programs which had gone undetected by many others, and to my father, who gave editorial advice. Finally, I must thank Sallie, Douglas, Andrew and Edward for their patience and encouragement.

Pawley's Island,
South Carolina,
June 4, 1976.

Contents

PREFACE   iii

1   Computer Systems and Spatial Data   1
1.1   Some Historical Background   2
1.2   Computer Number Systems   4
1.3   Computer Organization   6
1.4   Spatial Data   9
1.5   Scales of Measurement   9

2   Introduction to FORTRAN   11
2.1   The Assignment Statement   12
2.2   Variables and Constants   12
2.3   Arithmetic Expressions   13
2.4   Examples of Assignment Statements   14
2.5   Punching a FORTRAN Statement   14
2.6   Input and Output (READ, WRITE and FORMAT)   15
2.7   Some Special Input and Output Features   19
2.8   Example 1. DIST: Computing the Distance Between Two Points   19
      Exercises and Problems   24

3   FORTRAN for Spatial Data   25
3.1   DIMENSION Statements   26
3.2   Arrangement of Arrays in Computer Memory   27
3.3   DO Statements and Loops   27
3.4   DO Loops and Virtual Memory   29
3.5   Input and Output of Subscripted Variables   30
3.6   Example 2. MSDMAP: Mean and Standard Deviation of a Gridded Map   31
3.7   Conditional Execution (The Logical IF Statement)   34
3.8   Program Transfer   35
3.9   Flowcharting   35
3.10   Example 3. FREQ: Frequency Distribution Tabulation   37
3.11   Subroutine Subprograms   43
3.12   Adjustable Dimensions   44
3.13   Example 4. LISTR: Printing Spatial Data   45
3.14   LISTR with Exponential Format   50
3.15   Function Subprograms   51

3.16 Example 5. ARCDIS: Distance Between Points Given Their Latitudes and
     Longitudes   52
     Problems   54
     Suggestions for the Solution of Problems   55

4   Regularly Spaced Data with Implicit Coordinates   57
4.1 Map Coding into Matrix Form by Manual Techniques   57
4.2 The Grid Size   60
4.3 Character Data   62
4.4 Data Initialization   62
4.5 Example 6. MAP1: Line Printer Maps of Qualitative Data   64
4.6 Example 7. MAP2: Line Printer Maps of Quantitative Data   71
4.7 Example 8. MAP3 and MAP4: Line Printer Maps with Character Blocks   74
4.8 Example 9. SQUEZ1 and SQUEZ2: Preparing Matrix Data for MAP1
     and MAP2   80
4.9 Class Intervals on Maps of Quantitative Data   86
4.10 Example 10. MAP2A and SORT: Equal Areas of Grey Level Patterns
     on Quantitative Maps   87
4.11 Data Transformations   94
     Problems   96
     Suggestions for the Solution of Problems   96

5   Irregularly Spaced Data with Explicit Coordinates   98
5.1 Coordinate Systems   98
5.2 Transformation of Coordinates   99
5.3 Example 11. SCALE and POINT: Scaling and Plotting Points with a
     Line Printer   102
5.4 Isarithmic Maps from Irregularly Spaced Data Points   106
5.5 Two-Dimensional Interpolation   109
5.6 Example 12. INTERP: Line Printer Maps from Irregularly Spaced Data   111
5.7 Choropleth Maps from Irregularly Spaced Data Points (Proximal Maps)   115
     Problems   115
     Suggestions for the Solution of Problems   116

6   Lines and Networks   117
6.1 Regions as Polygons: Digitizing by Boundary Coordinates   117
6.2 Example 13. BDRY2: Filling a Matrix from Boundary Coordinates   119
6.3 Line Plotting Machines   126
6.4 Plotters and Spatial Data   128
6.5 Programming the Plotter   131
6.6 Example 14. CIRCLE: Graduated Circles   132
6.7 Networks   136
6.8 Example 15. SHORT: Minimal Distance in a Network   138
     Problems   141
     Suggestions for the Solution of Problems   141

7 **Some Programming Techniques**   142
7.1 Reducing Memory Requirements   142
7.2 Regularly Spaced Data Values and Irregularly Shaped Study Areas   147
7.3 Packaging Programs   149
7.4 Some Alternative Computer Languages: ALGOL, PL/I, BASIC, APL   152
    References and Selected Reading   154
    Appendix: The IBM 029 Card Punch   156
    Glossary   159
    Index   160

1
Computer Systems and Spatial Data

This book describes how computers are programmed to process and map spatial data. The person who reads it, studies the examples, and completes several of the assignments at the end of each chapter will have a good understanding of how a computer is used to solve spatial problems. He will be able to write his own programs for simple problems, and either select from existing programs for more difficult problems, or describe his requirements to a professional programmer.

A basic premise of this book is that computer programming must be learned, rather than taught, and that this learning is primarily achieved through running programs. It is assumed that the reader has access to a general purpose digital computer with a FORTRAN compiler, and that he will prepare, run, and debug modifications of most of the examples, and attempt several of the problems.

No previous computer experience is assumed, but this book does not dwell on the more elementary aspects of FORTRAN (or on the more advanced). Some readers, therefore, may wish to use a standard FORTRAN text as a supplement. Those with a background in a language other than FORTRAN are urged to use this book by translating examples and by solving problems in that language.

The book is organized as follows.

Chapter 1 presents some historical background, briefly describes the organization of a computer and reviews some concepts of measurement.

Chapter 2 is an introduction to the most elementary aspects of FORTRAN. The reader with computer background may wish to skim or to skip it.

Chapter 3 describes elements of FORTRAN required for handling sets of data values such as a matrix or vectors of coordinates. Considerable attention is given to virtual memory, an innovation in computer organization with significant consequences for spatial problems.

Chapters 4 and 5 discuss various techniques for mapping spatial data with the standard line printer available on all computers. Map coding is discussed and seven useful computer mapping programs are presented.

Chapter 6 includes three major topics: the coding of spatial patterns by the coordinates of region boundaries, the use of specialized line plotting equipment, and the representation and analysis of networks in computers.

Chapter 7 discusses several programming techniques which are particularly useful for problems involving spatial data.

A glossary defines technical terms and jargon.

1.1 Some Historical Background

The first large computers were built during the Second World War. Attempts were made earlier, the most remarkable being Charles Babbage's analytic engine. Babbage, an English mathematician and consulting engineer, designed in the 1820s a mechanical computing machine which would store a thousand fifty-digit numbers, add any two of these in one second, and use punched cards for entering new data values. (The punch card had been devised much earlier by the French inventor Jacquard to control weaving patterns on looms.) Babbage's scheme was not taken lightly; the British government invested several thousand pounds in the project, a considerable sum at the time. But even after 40 years, the analytic engine had not been completed, and the man and the idea died in 1871.

The prospect of this machine gave rise to the following comments by Lady Lovelace:[1]

It is desirable to guard against the possibility of exaggerated ideas that might arise as to the powers of the Analytical Engine. In considering any new subject, there is frequently a tendency, first to overrate what we find to be already interesting or remarkable; and, secondly, by a sort of natural reaction, to undervalue the true state of the case, when we do discover that our notions have surpassed those that were really tenable.

The Analytical Engine has no pretensions whatever to originate anything. It can do whatever we know how to order it to perform. It can follow analysis; but it has no power of anticipating any analytical relations or truths. Its province is to assist us in making available what we are already acquainted with.

Lady Lovelace showed perception. Today, with computers essential to our way of life, her observations are perhaps even more appropriate. They emphasize the process of programming and the importance of the programmer's knowledge, and point out the role of programs and computers—to serve man.

After Babbage, there were few significant developments in machine computing until 1940. Most important, perhaps, was Dr Herman Hollerith's use of punched cards in the 1890 US Census. Also significant was the construction of several specialized machines in the 1920s and 1930s for calculating ballistic trajectories and gunnery control. In the decade starting in 1939, three fundamental advances took place. The first of these was the construction of the Mark I Automatic Sequence Controlled Calculator at Harvard University under the direction of Dr Howard Aiken and with the support of International Business Machines. This was an electromechanical machine using several thousand electrical relays. It was quite slow, taking four and one-half seconds to multiply two 23 digit numbers.

Even before the Mark I was completed in 1944, work had started on the first electronic computer at the Moore School of Electrical Engineering of the University of Pennsylvania. By using vacuum tubes instead of electrical relays, it was possible to speed computing by a factor of 1000. The machine was called ENIAC—Electronic Numerical Integrator and Calculator—and was designed by J. Presper Eckert, an electrical engineer, and John Mauchly, a physicist. It was completed and put in operation in 1946.

The third important advance of the 1940s was the concept of the stored-program

[1] From Jeremy Bernstein, 1966, *The Analytical Engine*, New York: Vintage Books, pp. 44–5.

computer. In the Mark I and ENIAC, numbers were stored in a memory, and instructions about what was to be done with these numbers were represented by electrical circuits. If a different set of operations was required, the electrical circuits had to be rewired, a difficult and time-consuming job. The principle of a stored program computer is that some of the numbers in the memory are actually codes for instructions; they are the equivalent of the wires on earlier machines. The computer operates by examining a number representing an instruction, making the necessary switching arrangements (in effect, doing the wiring), and then executing the instruction with a data value. Thus there is a sequence of instruction cycles and execution cycles. With such a computer, the need for rewiring is eliminated; a program of instructions is read by the machine from punched cards (or similar devices) along with the data which are to be processed.

The concept of the stored-program computer was developed at the Moore School, primarily by John von Neumann, Herman Goldstine and Arthur W. Burks. Even before ENIAC was completed, this group had started the design and construction of a stored program computer called EDVAC, or Electronic Discrete Variable Automatic Computer. The machine was completed in 1950, but by then the first stored-program computer was already in operation at Cambridge, England—EDSAC (Electronic Delay Storage Automatic Calculator), built under the supervision of M. V. Wilkes.

In 1951, the first commercial stored-program computer was delivered to the US Bureau of the Census. This machine (UNIVAC-I, Universal Automatic Computer) was built for Sperry Rand by Eckert and Mauchly, the originators of ENIAC. During the two decades since then, many thousands of computers of many sizes and prices have been built by many firms, with astonishing improvements in speed and reliability.

Even given the advantages of stored programs, programming for early computers was an extremely tedious business because each operation required so many instructions, and everything had to be stated exactly in machine code. This was such a limiting factor to computing that very early in the 1950s, there were attempts to write translating programs which would read statements resembling arithmetic, then compile these into machine language instructions. (These programs are today termed compilers.) One of the first of these was started in 1954 at IBM by John W. Backus, and was released to the public in 1957. It was called FORTRAN (*formula trans*lator) and has since become the predominant scientific programming language in the world.

Since the introduction of FORTRAN, hundreds of computer languages have been written, but most are either very specialized, or are little used because they are inefficient, poorly documented, or otherwise unsuitable. Only two general purpose languages other than FORTRAN have survived and are widespread: ALGOL (far more important in Europe and the UK than in the US or Canada), and PL/I (Programming Language One). These languages will be briefly described in the last chapter of this book, together with some more recently developed languages (BASIC and APL). FORTRAN will be described in some detail and used in the examples throughout this book.

The application of computing to spatial problems commenced in the early 1960s in several universities in the United States, notably the University of Washington, the University of Chicago, Northwestern University, and the University of Michigan. By 1965, considerable research was underway in the geography departments of these and several other larger institutions in which computers were an essential data processing tool. Many graduate students were acquainted with FORTRAN and packaged programs, and a number were proficient programmers.

Since 1965, a computer capability has become a standard part of university programs in geography and planning throughout the world. This typically consists of access to a university computer, a collection of programs obtained from other institutions (or more recently, through the Geography Program Exchange[2]), and a small number of locally written programs. These programs are generally of three types: mapping, statistical, and optimizing. The program library is usually supervised by a graduate student or, less often, a scientific programmer or faculty member.

Computers are now used in some way in most empirical research in geography. All graduate students are expected to have some sense of the appropriate use of computation, and many graduate students and undergraduates learn a computer language, write elementary programs, and run larger programs with data they have acquired. This book is written primarily for these people.

1.2   Computer Number Systems

Binary organization is a fundamental element in the design of digital computers.[3] All data and instructions are represented as the presence or absence of electronic charges (conventionally represented as a pattern of zeros and ones). Even the size of computers may be expressed in binary; the number of storage locations in memory, for example, is always a power of 2.

The reasons for this binary operation and organization are due to engineering characteristics and need not normally concern the programmer since decimal numbers, symbols, and instructions are automatically converted to their binary equivalents by the machine. The only contact the typical user will have with the binary number system is when considering the size of the individual storage unit. In IBM computers, storage locations are described in terms of bits, bytes, and words. A bit is the smallest single element in the machine, and can have only a 0 or 1 value. A byte is made up of eight bits, and a word is four bytes. Most users operate only with words. Under special circumstances, one may choose to operate with half-length (two bytes) or double-length words (8 bytes), but it is usually difficult to deal only with single bits (depending on the language).

The correspondence between decimal and binary numbers is shown in Table 1.1 for values up to 16. Examination of this table will give one the rules for forming a sequence of binary numbers. The procedure involves adding ones until all positions are filled, then adding a position to the binary number. Also shown in the table are corresponding hexadecimal numbers. This is a number system with base 16 which is

---

[2] Geography Program Exchange, Computer Institute for Social Science Research, 515 Computer Center, Michigan State University, East Lansing, Michigan, 48823, USA.

[3] A distinction is usually made between digital and analogue computers. In the latter, values are stored as the strength, amplitude or intensity of an electrical signal; in the former, values are stored as numbers coded with a system described below. Analogue computers are usually special purpose machines used for particular types of mathematical calculations.

Table 1·1   Number Systems

Decimal	Binary	Hexadecimal
0	0	0
1	1	1
2	10	2
3	11	3
4	100	4
5	101	5
6	110	6
7	111	7
8	1000	8
9	1001	9
10	1010	A
11	1011	B
12	1100	C
13	1101	D
14	1110	E
15	1111	F
16	10000	10

often used as a shorthand method to represent strings of binary bits and sometimes to state the size of core storage. The typical user will rarely need to translate these values into decimal, if, in fact, he even encounters them. If conversion is required, tables are available to simplify the task.

There is obviously an upper limit to the size of a number which can be stored as a string of binary bits. With a 32 bit word (the IBM 360/370 standard), this limit is between 7 and 8 digits. To allow for larger numbers, an alternative method of storage may be used which converts a number into two parts, a mantissa and an exponent. For example, the number 432.689 may be represented as the product of 0.432689 and 10^3. In this case, 0.432689 is the mantissa and 3 is the exponent. Both mantissa and exponent are still stored in a single computer word,[4] but the new format means that much larger numbers may be stored, generally up to 10^{99}. This storage method is termed floating point, and is used for most numerical computations. The first method is termed fixed point, and is normally used only for integer numbers. (This distinction will be discussed further in Chapter 2.)

Non-numeric information such as alphabetic characters may also be stored in computers. The principal convention for this is EBCDIC (Extended Binary Coded Decimal Interchange Code). In this system, a single character is represented by a pattern of eight bits in one byte (thus a maximum of four characters in a standard 32 bit computer word). Eight bits allow a total of 256 characters to be defined, so there is no problem accommodating the alphabet, numerical characters, and many special symbols. Non-numeric data are discussed in detail in Chapter 4.

[4] The *significance* of a number stored in this way is not different from one stored directly as a string; only the first seven digits of the number may be stored as mantissa.

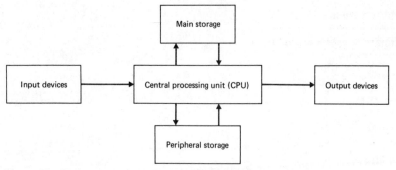

Figure 1.1   Basic components of a general purpose digital computer.

1.3   Computer Organization

General purpose digital computers range in size from desk-top models to systems which fill several rooms with machinery. All have the components illustrated in figure 1.1 in common.

Input devices are machines through which the user gains access to the computer. The most common of these are readers for punched cards. The conventional punch card has 80 vertical columns and rows marked 0 to 9 (figure 1.2). A single numerical digit is represented by a slot punched in the appropriate row number. Alphabetic or special characters are represented by a combination of one or more digit slots, plus a slot in the upper part of the card (where row numbers are not indicated). Cards are punched using machines such as the IBM 029 card punch (plate 2); the operation of this machine is described in the Appendix. Card readers are able to scan cards at rates up to several hundred per minute, and are so reliable that it is common practice to allow users to operate them with no special instructions.

Far slower but considerably more convenient for some problems are remote terminals resembling typewriters or TV consoles (termed cathode ray tubes or CRT'S). With these devices, the user is able to communicate directly with the computer, not only to send instructions, but also to receive messages or results in seconds or less. The input and output functions are thus combined in the same machine, although

Figure 1.2   A typical punch card (80 columns).

Plate 1 An IBM 370/168 computer. This is one of the larger systems in use at present. The central processing unit and main storage are located in the cabinets at the rear. The peripheral storage units are located along the left and right walls. A card reader is in the left foreground, and two printers in the right foreground. The operator's console is in the center. (Photograph provided by IBM.)

both are so slow that these devices are normally used only for special purposes such as debugging a program or retrieving a single item of information from computer storage.

The central processing unit or CPU contains the electronic circuits and registers that control the sequence of computing steps, do the arithmetic operations and monitor the flow of numbers and information into, out of, and within the machine. Computer charges are usually largely based on the amount of time a program actually uses the CPU.

The principal difference between large and small computers is not speed of operation, but the amount of main and peripheral storage. A very large machine such as an IBM 370/165 will usually have space for over a million numbers in main storage, any one of which can be transmitted to the CPU in an instant. Peripheral storage in a large machine consists of magnetic discs, tapes, or drums which have capacities of hundreds of millions of numbers. These data are not as accessible as those in main storage, however, since the peripheral device must first be positioned at the desired location, and then the information read and transmitted to the CPU registers.

The standard output device is a line printer, a machine which prints an entire 132 character line at once, at a rate between 300 and 1100 lines per minute (Plate 3).

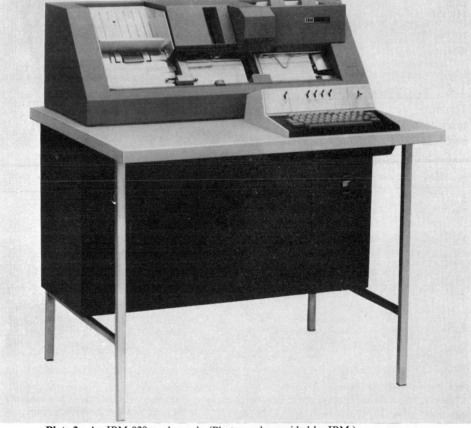

Plate 2   An IBM 029 card punch. (Photograph provided by IBM.)

Plate 3   An IBM 3211 high speed printer. (Photograph provided by IBM.)

Some newer electrostatic machines print several different type styles at rates three to four times faster than mechanical line printers, and do line plotting as well. Unfortunately, there are very few such devices yet in operation. Other graphic output devices are described in Chapter 6.

1.4   Spatial Data

To complete this introductory chapter, we will now briefly consider categories and properties of spatial data in order to introduce and define the terms which will be used later.

Spatial data occurs in three forms:
1   Aerial photographs, remote sensing, or general-purpose topographic maps which require some degree of interpretation to determine distributions of single factors.
2   Lists or tabulations of values which occur at specific places (at coordinate locations, or within zones), such as climatic data for weather stations, topographic elevations of a set of points, locations of a particular plant species or populations of counties.
3   Thematic maps which show the distribution of single factors such as geology, vegetation, population density or land use.

1.5   Scales of Measurement

Whether the spatial data are in the form of a map, an aerial photograph, or a list, they are scaled or given values in some way. The location information is either presented directly as numerical coordinates (in a list), or can be measured in this form from a map. The data values, however, are presented or can be measured in a variety of ways, as often in a non-numerical format as with numbers. The method by which the data are scaled is critical because it determines what programming techniques may be used and whether, in fact, certain operations are even possible in a computer.

Four kinds of measurement scales are generally recognized: nominal, ordinal, interval, or ratio. The first two are known as qualitative or non-parametric scales, the latter as quantitative or parametric scales. A description of each type of scale follows.

1   Nominal   Commonly all that is known about a particular place, area, or observation is that it belongs to a set of similar objects which is different from others. One can describe these differences in terms of the presence or absence of properties, but it is impossible to state by what degree one set differs from another. Some examples of nominal data are rock types, landforms, vegetation classes, and land use classes. Nominal data values are always given as labels or names. Numbers may be used, but they have meaning only as labels. (In fact, it will become apparent in Chapter 4 that using numbers as labels for nominal data greatly simplifies many computer operations.) Arithmetic operations such as addition and subtraction have no meaning with nominal data. The only numerical operations possible are counting occurrences or measuring area or perimeter.

2   Ordinal   Often one is able not only to classify objects or places into sets, but is also able to rank the objects or sets according to some criterion. An example commonly found in geography is ranking urban centres by their size (population). In general, arithmetic operations may not be used with ordinal data because the difference between ranks is unknown. In a set of 20 ranked centres, for example, we have no idea whether the interval between the first and second is even similar to that between the nineteenth and twentieth. Counting occurrences and measuring areas is possible, as with nominal data, along with some specialized operations (non-parametric statistics and multidimensional scaling). Few spatial data sets occur in the ordinal form unless they have been transformed for a particular purpose.

3 Interval When one is able to state by how much one object or place differs from others, the data are said to be parametric. On an interval scale, the zero point and the value of a unit are arbitrary. Centrigrade and Fahrenheit temperature scales are good examples of interval scales. Any arithmetic operations may be applied to interval data, but the results may not always have meaning. For example, one might attempt to compare day and night temperatures in a place by computing the ratio between them, but this ratio has no real meaning other than stating the relationship between two numbers. If the day's maximum is 10°C and the night's minimum 5°C, the ratio between them is 2.0, but it would be absurd to state that the day was twice as warm as the night.

4 Ratio If one knows not only how objects or places differ from one another, but also from a zero point which has real meaning, the data are said to be on a ratio scale. Some examples are the Kelvin temperature scale, distance, longitude and latitude, reflectance from a photograph (grey level), population, income, production, and employment. All arithmetic operations may be used with such data.

2

Introduction to FORTRAN

There are dozens of dialects of the FORTRAN language. Each major computer manufacturer usually has a version particularly suited to its type of machine, and often this dialect exists in several forms reflecting extensions in the capability of the language or modifications for particular machines. This book will not deal with these variations, but will present a version of FORTRAN acceptable to most large machines. It consists of USA Standards Institute FORTRAN IV plus several extensions which are commonly available.[1]

Table 2.1   The FORTRAN vocabulary

Word	Estimated frequency of use (1 = most often; 3 = least often)
ASSIGN	3
BACKSPACE	3
BLOCK DATA	3
CALL	1
COMMON	2
COMPLEX	3
CONTINUE	1
DATA	2
DIMENSION	1
DO	1
DOUBLE PRECISION	3
END	1
END FILE	3
EQUIVALENCE	3
EXTERNAL	3
FORMAT	1
FUNCTION	1
GO TO	1
IF	1
INTEGER	2
LOGICAL	3
READ	1
REAL	2
RETURN	1
REWIND	3
STOP	1
SUBROUTINE	1
WRITE	1

Table 2.1 lists the FORTRAN vocabulary used in this book, together with the author's estimate of the frequency of use of each word for the types of problems encountered with spatial data. Perhaps the most significant thing about this table is its

[1] The criteria were that the extension had to be particularly useful, available on IBM machines (the most common), and available on most other large machines as reported in Stuart (1970).

brevity; the total vocabulary of the language is about thirty words (forty in the most extended versions). Each word, however, has a well defined meaning and can be used only in specified ways in certain types of statements. There is no allowance for error; a mistake, even in spelling, will generally mean that the program will not be executed by the machine, but returned with an error message.

It is obvious that learning FORTRAN is much simpler than learning a foreign language. This is because there is no vocabulary problem, the rules are clearly stated with no exceptions, and the computer will identify and reject most errors. (In some cases, as with the WATFIV compiler, an attempt is made to diagnose the error and to suggest corrective action.) There is a similarity between FORTRAN and a foreign language, however, in that knowledge of the mechanics is only a preliminary to using the language well. Someone who has mastered FORTRAN may still not be a good computer programmer.

In this chapter, six basic FORTRAN statements are presented (assignment, READ, WRITE, FORMAT, STOP and END), and their use illustrated with an example program. In the following chapter, several additional statements are presented which are required for spatial data and more advanced programming problems.

2.1   The Assignment Statement

The FORTRAN assignment statement is sometimes described as an arithmetic statement because of its resemblance to an identity or equation. Some examples of assignment statements are

```
A=B+C*X
C=(X/Y)/D
RESID=XAB**2
DMAX=1.874*AMAX
```

Each of these four statements is an instruction to the machine to evaluate the expression on the right side of the equals sign, and store the result as the new value of the quantity on the left side of the equals sign. It is important to realize that these statements are not arithmetic identities or equations because

```
A=A+B
```

is an acceptable statement in FORTRAN although it is mathematically absurd. The equality symbol must be interpreted as a replacement operation, an instruction to store a new value at an addressed location in the computer memory. For this reason, only one name will ever appear on the left side of an assignment statement.

The assignment statement is subject to a number of rules in its formation. These involve the basic units of the statement (variables and constants) and the way these elements are combined on the right side of the statement (arithmetic expressions).

2.2   Variables and Constants

The numbers which are manipulated by a computer program are either integer (also termed fixed point) or real (also termed floating point or decimal). The distinction is important in programming because each type is operated on and stored differently in the computer,[2] and read in and printed out with different conventions. The basic

[2] Real numbers are represented in the machine by a mantissa and an exponent, in the same manner as logarithms. Integer numbers are represented in their full form.

difference between the two types is that integer numbers do not have a decimal point, but real numbers do. Thus 2, 97, 1034, and 6102 are integer numbers, and 2.0 11.2, 4.3291 and 0.63 are real or floating-point numbers. (There is a difference to the machine between 3 and 3.0.)

This convention poses no particular problem with constants, since a decimal is or is not punched with the number depending upon what is appropriate. In the case of variables, the distinction appears in naming, in that integer variables must begin with *I, J, K, L, M,* or *N,* and real variables with any letter other than these six. (Ways to override this rule are presented later.) Any (or no) alphabetic or numeric may be used to fill out the variable name (up to a maximum of six characters), but no special characters such as +, /, −, or *.

Some examples of correct variable names are:

```
ANSWER
GEORGE
XX
HOLD
JUMP
KK4
```

The first four are real names, the last two integer. Two examples of incorrect variable names are

```
ANSWERS
GO+1
```

The first of these has more than six characters, the second contains the illegal character +.

2.3   Arithmetic Expressions

Usually one does not find a single variable or constant on the right side of an assignment statement. Two or more of these elements are usually combined into an arithmetic expression by means of the arithmetic operators:

+ addition
− subtraction
* multiplication
/ division
** exponentiation

One of these operators must always separate variables or constants from one another. (This is why the arithmetic operator symbols may not be used as part of a variable name.)

It should be apparent that ambiguities may sometimes arise in arithmetic expressions. For example, if it is desired to evaluate

$$\frac{b+c}{d}$$

the FORTRAN arithmetic expression

```
B+C/D
```

actually means

$$b+\frac{c}{d}$$

This is because a hierarchy of operations exists in the computer. Exponentiation operations always take precedence, followed by multiplication and division, then addition and subtraction. This means that in the example above C is divided by D first, then the result added to B. If the expression were

    B+C/D**2

then D would be squared first, then divided into C, and the result added to B. Note that multiplication has the same precedence as division, and addition the same as subtraction. If two or more operators with the same precedence occur in an expression, the operations are evaluated from left to right.

This built-in hierarchy of operations can be overridden by using parentheses. Any expressions within parentheses are computed first. Thus in the expression

    (A+B)*(A-B)

the quantities within the parentheses will be evaluated before the multiplication, but in

    A+B*A-B

A and B will be multiplied together, then the addition and subtraction will take place.

2.4   Examples of Assignment Statements

Some representative FORTRAN assignment statements are presented below together with a description of their meaning.

A=B	The value of A is replaced by the current value of the real variable B.
X=6.413	The value of X is replaced by 6.413.
PQ=PQ+1.0	The value of PQ is replaced by a value 1.0 greater.
X=A+B*4.6	The value of the real variable B is multiplied by 4.6, the result added to the current value of A, and the result of this replaces the value of X.
ALPHA=(A-B)**0.5	The current value of B is subtracted from that of A and the result raised to the 0.5 power (the same as taking the square root); this replaces the value of ALPHA.
IJK=R*4.612	The product of the current value of the real variable R and 4.612 is truncated then replaces the value of IJK. (By definition, decimal places must be removed before a number can be stored as an integer. In assignment statements this is done by truncation, that is, the dropping of all digits after the decimal place without rounding off—3.99999 becomes 3, not 4.)
R=IJK*4.612	The product of the current value of the integer variable IJK and 4.612 replaces the value of R. (There is no question of truncation in this case, the reverse of that just presented.)

2.5   Punching a FORTRAN Statement

Assignment and other FORTRAN statements are punched one statement to a card, and only in columns 7 to 72 (an important and easily forgotten rule). Columns 1 to 5 are reserved for statement numbers (to be discussed later). Column 6 is used to indicate that the information on this card is a continuation of that on the preceding card. If any number or character is punched in this column, the compiler links the information from columns 7 to 72 on this card with the FORTRAN coding on the previous card. Anything punched in columns 73 to 80 will be transmitted by the card reader and printer, but will be ignored by the FORTRAN compiler. This space is usually left blank or is punched with card sequence numbers (so that the program deck can be easily re-assembled in case of an accident). Figure 2.1 illustrates a typical FORTRAN statement card. Note that the card printing is designed as a reminder of these conventions about punching, and also to simplify checking a program before submitting it to the computer.

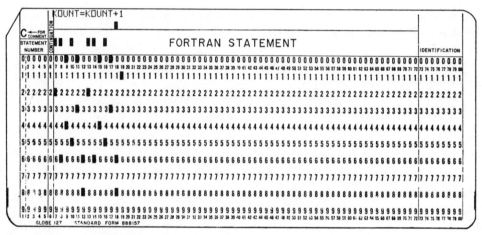

Figure 2.1   A typical FORTRAN statement card.

2.6   Input and Output (READ, WRITE and FORMAT)

Three FORTRAN statements are used to transfer data into and out of the computer. Two of these (READ and WRITE) are executable instructions or commands, the third (FORMAT) is not executable, but describes how data are arranged on a card or on a page of output.

READ statements are commands to read values for specified variables from data cards. The general form is

READ (*m, n*) VAR1, VAR2, VAR3

where VAR1, VAR2, and VAR3 represent the names of variables whose values are being read. The information within parentheses indicates that: (a) the data card is stored in a data set labelled *m*; and (b) FORMAT statement number *n* contains the necessary information about how the data are punched on the card. The data set number depends on the manufacturer, local conventions, and input procedure being used. For IBM machines where the data are on standard cards, the data set reference number is 5. (Since this number is the same for most users, some compilers allow the user to delete it; see the following section.)

The symbol *n* in the READ statement is the number of a FORMAT statement somewhere in the FORTRAN program which describes how the numbers are punched on the data card(s). This statement *always* has a number somewhere in columns 1 to 5 on its card, followed by the word FORMAT starting in column 7 or later. Coding is placed within parentheses to describe the exact arrangement of the numbers, in general form

n FORMAT (S1, S2, S3, . . ., Sm)

where n is a statement number which appears in a READ statement (or a WRITE statement, described below), and S1, S2, etc. are format field specifications. The code for field specification varies with the kind of data: real, integer, and other types yet to be described all have different specification codes. Integer or fixed-point data are specified by a field of the form

Iw

where w is the number of spaces or width occupied by the integer number field. Real or floating-point data are defined by

$Fw.d$

where w is the number of spaces occupied, and d is the number of places to the right of the decimal point.

An example of a READ and FORMAT statement pair is

```
     READ(5,9)A,JKL,BETA
9    FORMAT(F5.1,I6,F7.2)
```

Figure 2.2 is an example of a data card which could be read by these statements. A value for A is punched in columns 1 to 5, a value for *JKL* in columns 6 to 11, and a value for BETA in columns 12 to 18. In all cases, the number is right-justified within the specified field.

Figure 2.2   A data card

It is very common when several numbers are punched on a card that the same field specification can be used for each. For example, seven floating-point numbers might be punched on a card in fields of five columns with the decimal punched in the fifth column in each case. The FORMAT statement for this card would be

```
12   FORMAT(7F5.0)
```

The 7 before the field specification indicates that it is to be repeated seven times.

The WRITE statement is an instruction to print the current value of a variable or variables. It is of the same general form as the READ statement:

WRITE (*m, n*) VAR1, VAR2, VAR3

where VAR1, VAR2, and VAR3 are variable names and *n* the FORMAT statement number. The data set reference number *m* is usually equal to 6. FORMAT statements used with a WRITE statement are slightly different from those used for reading in order to include information about carriage control, or how output lines are to be spaced. For single spacing, the part of the FORMAT statement within parentheses is started with ' ', for example

```
        WRITE(6,4)ALPHA
4       FORMAT(' ',F6.1)
```

The other carriage control symbols are[3]

'0' double spacing

'−' triple spacing

'1' start a new page

'+' no vertical spacing (print next line in same place)

If the carriage control symbol is omitted from an output FORMAT statement, the compiler uses by default the first character or number in the output line. In most cases, this will be a blank, and single spacing will result.

It is often desirable or necessary to space number fields well apart when printing, or to skip certain parts of a data card when reading. This can be done by inserting a special field in the FORMAT statement where the spacing is desired. This field has the form

*n*X

where *n* is an integer constant equal to the number of columns to be skipped (if reading) or the number of blanks to be printed (if writing). A case where the X format code is necessary is when a data card is read which has numbers on it which are not required. For example, assume that there are ten numbers on a card, punched 10F5.0. When the program is written which uses these data, it is decided that the third and ninth numbers are not required. This could be done as follows:

```
        READ(5,962)A,B,C,D,E,F,G,H
962     FORMAT(2F5.0,5X,5F5.0,5X,F5.0)
```

No X field is necessary when the numbers to be skipped occur after the last number to be read. On the card described above, it may be desired that only the first seven fields be read:

```
        READ(5,76)A,B,C,D,E,F,G
76      FORMAT(7F5.0)
```

When the READ statement is 'satisfied', execution then proceeds to the following coding.

Some examples of READ, WRITE and FORMAT statements are presented below together with a description of their meaning. The upper part of the corresponding data card is shown for READ statements.

---

[3] These are the most commonly used characters; some machines differ, and some do not offer such features as triple spacing or no vertical spacing.

```
      READ(5,12)A,P,IJK,ALPHA
12    FORMAT(2F5.0,I6,F5.1)
```

Values for four variables are read. The first two and the last one are real, the third is integer.

```
      READ(5,422)GAMMA,DELTA,NEXT
422   FORMAT(F6.2,F4.0,5X,I7)
```

Values for three variables are read. After the first two, the FORMAT specifies that five columns be skipped. (If any numbers or characters were punched here, they would be ignored in the READ operation.)

```
      READ(5,16)A,B,C,D,E,F,G,H
16    FORMAT(5F5.0)
```

Values for eight real variables are read. However, the FORMAT statement only allows five data fields per card. In this case, the values for the last three are punched on a second card and the READ statement re-uses the FORMAT statement until it is satisfied, i.e. until values for all eight variables have been read.

```
      WRITE(6,93)
93    FORMAT('1')
```

The printer is positioned at the top of a new page. No values are printed.

Punching errors in READ, WRITE or FORMAT statements are readily detected and rejected by most FORTRAN compilers. However, some errors are accepted by the compiler, execution attempted, and often completed, but with strange or un-expected output. Three errors of this type are as follows:

1 Mismatch of the format codes and fields on a data card, so that parts of two different fields are read as the value of a variable.
2 Omission of the carriage control field from an output FORMAT.
3 Specifying too long a card or line. (The standard data card length is 80 columns, and printer line 132 characters.)

2.7 Some Special Input and Output Features

Input and output in FORTRAN is the most tedious part of the language, and probably the principal source of errors. For this reason, some FORTRAN compilers have been extended to allow simplified input and output procedures in addition to those in standard FORTRAN. Some versions, for example, free the user from the need to specify the data set reference number, thus:

READ *n*, VAR1, VAR2, VAR3

and

PRINT *n*, VAR1, VAR2, VAR3

where *n* is a FORMAT statement number. There are some compilers which go so far as to eliminate the need for the FORMAT statement, using one or more blanks between numbers on cards as a signal of a new field, and printing in a standard format. (IBM FORTRAN'S NAMELIST is an example.) The reader is advised to investigate whether any such extensions are part of the compiler he is using, because they are very useful for simple problems. However, for problems involving spatial data, particularly when map-type output is desired, these simplified procedures are generally not adequate.

2.8 Example 1. DIST: Computing the Distance Between Two Points

We have now described the assignment statement, that FORTRAN statement which actually specifies the computing to be done, and READ, WRITE and FORMAT statements which instruct the computer about data which are to be read and results which are to be printed. These statements are sufficient to code elementary programs when they are combined with system control cards which are specific to each installation and two additional FORTRAN statements which are required to stop a program.

The two FORTRAN statements which signal the end of a program are STOP and END. STOP is almost always required, but it is possible to use other techniques. END is always required, and must always be the last card in any FORTRAN program. The difference between the two is that STOP actually instructs the computer to stop execution, while END is a signal that there are no more FORTRAN statements. STOP is thus an executable statement, while END is non-executable.

System control cards are required for identification, accounting, and instructions about how a computer program is to be executed. These cards can be complicated; in fact, they are usually considered to be a language themselves (such as IBM's Job Control Language). Fortunately, computer centres have catalogued the most common of these procedures so that the programmer need only prepare relatively simple cards; the machine's operating system automatically generates the additional system information required. Fast compilers such as WATFIV have simplified system card requirements even further.

The distance between two points, given their coordinates, is an application of the Pythagorean theorem: the length of the hypotenuse of a right-angled triangle is equal to the square root of the sums of squares of the other two sides of the triangle. Thus, if the coordinates of the two points are (x_1, y_1) and (x_2, y_2) (Figure 2.3), the distance between them is

$$\sqrt{(x_1 - x_2)^2 + (y_1 - y_2)^2}$$

This relationship may be expressed in a single assignment statement such as

`DIST=SQRT((X1-X2)**2+(Y1-Y2)**2)`

The tricky problem of how to compute square roots is handled here with the FORTRAN-supplied function SQRT which automatically computes the square root of the quantity in parentheses which follows it. (More such functions are described in the following chapter.)

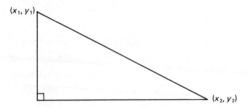

Figure 2.3   The distance between two points.

Other statements required for this program are those which read values for the coordinates of the two points, print the answer, and stop the program. Not required, but usually included in all computer programs are comment cards. Any card in a FORTRAN program deck which has a C punched in column 1 is ignored by the compiler, but is printed on the program listing. These cards are very useful for identifying various parts of a program and describing the method of computation. The novice programmer should use comment cards freely.

The required FORTRAN statements are arranged into a program by placing the executable statements (assignment, READ, WRITE and STOP) in the order in which these operations are to be carried out by the machine. The FORMAT statements, which are non-executable, may be placed anywhere in the program, but are usually either placed immediately after the READ or WRITE statement which refers to them, or collected in a group at the end of the program. The END statement terminates the FORTRAN program, and comment cards are inserted wherever they seem appropriate.

Before the program is punched, it is written by hand either on ordinary paper, or on coding forms which have been specially designed for FORTRAN statements and data. Figure 2.4 is an example of such a coding form. Note that the alphabetic character O is distinguished from the numeral 0 by a slash through it, and I and 1 are both carefully formed. These distinctions are essential if the coding form is to be punched by someone other than the programmer. (Unfortunately, there is not complete agreement about these conventions; it is becoming increasingly common to put the slash through the zero.)

A listing for the program is presented below in which an acronym (DIST) and a sequence number appears on the far right of each statement. The purpose of the information in this book is to identify those FORTRAN statements which are part of an example program. Any statement which does not have such information on the far right is an example of a feature of the FORTRAN language or an illustration of a programming technique, but not part of one of the examples. (When a program has been thoroughly debugged, however, it is common to add a name and number to each card so that the deck may be easily re-assembled in case of accident.)

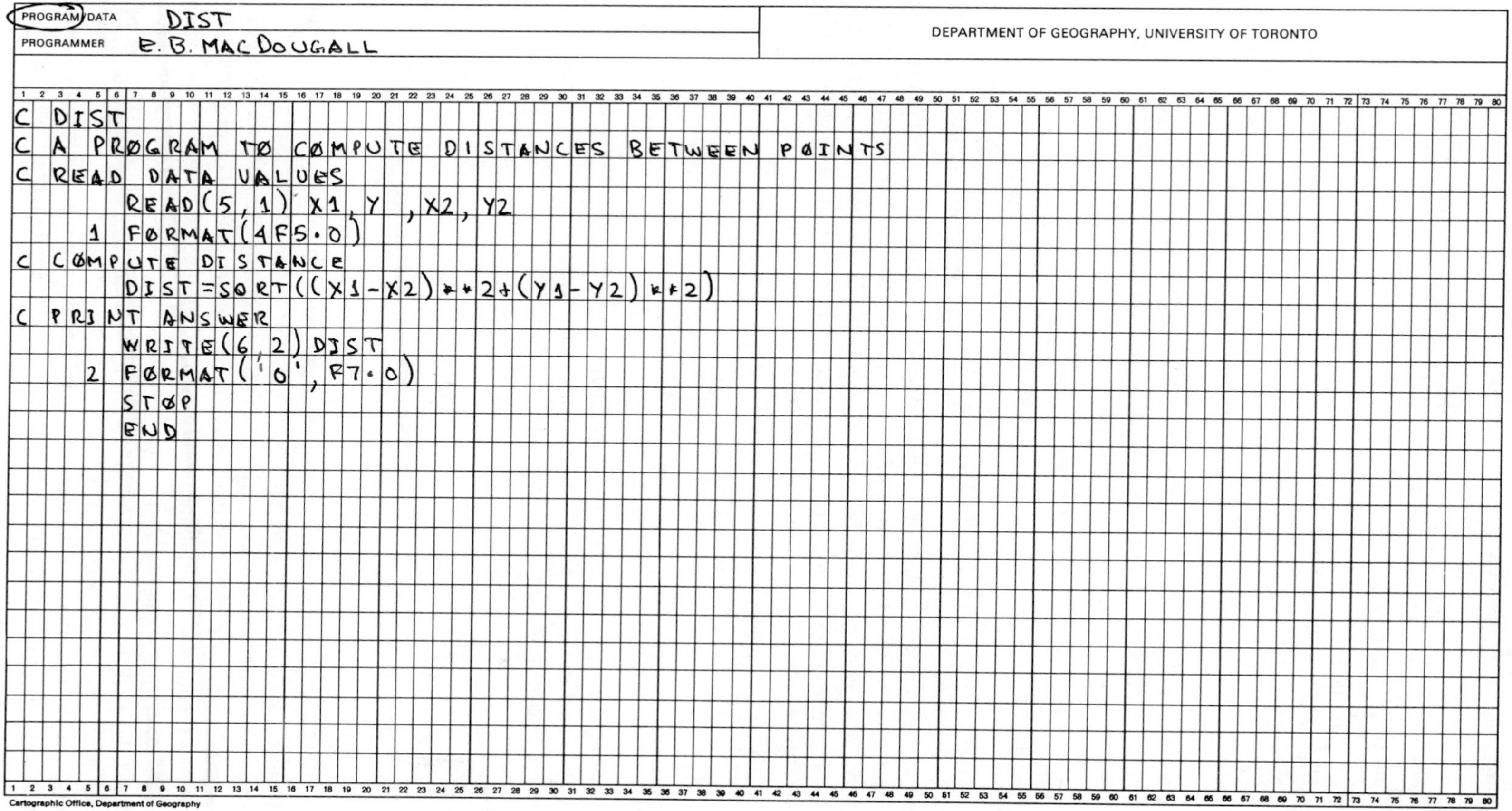

PROGRAM/DATA DIST

PROGRAMMER E. B. MAC DOUGALL

DEPARTMENT OF GEOGRAPHY, UNIVERSITY OF TORONTO

```
C  DIST
C  A PROGRAM TO COMPUTE DISTANCES BETWEEN POINTS
C  READ DATA VALUES
      READ(5,1) X1, Y1, X2, Y2
1     FORMAT(4F5.0)
C  COMPUTE DISTANCE
      DIST=SQRT((X1-X2)**2+(Y1-Y2)**2)
C  PRINT ANSWER
      WRITE(6,2) DIST
2     FORMAT('0', F7.0)
      STOP
      END
```

Cartographic Office, Department of Geography

Figure 2.4 A coding form.

```
C DIST                                                  DIST   1
C A PROGRAM TO COMPUTE DISTANCES BETWEEN POINTS         DIST   2
C READ DATA VALUES                                      DIST   3
      READ(5,1)X1,Y1,X2,Y2                              DIST   4
    1 FORMAT(4F5.0)                                     DIST   5
C COMPUTE DISTANCE                                      DIST   6
      DIST=SQRT((X1-X2)**2+(Y1-Y2)**2)                  DIST   7
C PRINT ANSWER                                          DIST   8
      WRITE(6,2)DIST                                    DIST   9
    2 FORMAT('0',F7.0)                                  DIST  10
      STOP                                              DIST  11
      END                                               DIST  12
```

A data card which gives values for X1, Y1, X2, and Y2 must be provided with the FORTRAN program. An example of a suitable card containing four values in the format specified in statement 1 of the program is shown below:

These data could be punched on a card in two other ways, and yet have the same values read into the machine by the same program. In the first of these, shown below, no decimals have been punched. The FORTRAN compiler will automatically insert a decimal in the appropriate location, in this case, after the last digit of each field. In a sense, the machine is inserting decimals between card columns.

In the second alternative, the data have been punched in fields of F5.1, rather than the F5.0 specified in the FORMAT statement. In such cases the punched decimal on the card rules, overriding the specifications in the program.

The FORTRAN program deck and the data card are combined with those system cards required for a particular compiler and computer to form what is termed a job or a job deck. Figure 2.5 shows the entire deck for DIST laid out with the WATFIV system cards currently required at the University of Pennsylvania. The programmer then submits the job to the computer, either by handing it to an operator who places it in a card reader, or by actually operating a card reader himself. He then waits for a period of time, depending on backlog, priority class, and conventions at the particular installation.

Eventually the programmer is returned several pages of output containing system information (usually unintelligible to the novice, but eventually useful), a statement of

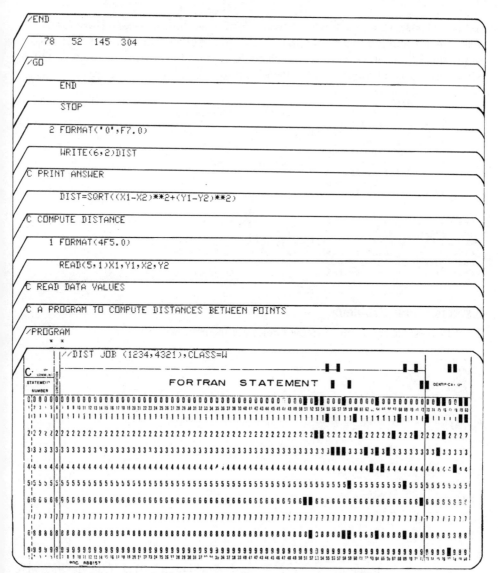

Figure 2.5   The cards for DIST.

charges, a listing of the FORTRAN program he submitted, and either the output he expected from his WRITE statements, or error messages from the FORTRAN compiler. The beginner will more often receive error messages than correct output. On FORTRAN compilers designed for production (such as IBM's FORTRAN H), error messages are so cryptic that they are useless to the novice. On compilers intended for program debugging and testing (such as WATFIV), the user is not only told that he made an error, but is warned of possible errors, and has obvious mistakes automatically corrected. For example, below is a listing of DIST which was submitted with two errors: the FORMAT statement was numbered incorrectly, and the word END was spelled EMD. The WATFIV compiler caught the statement number error (the last two messages), and detected a missing END (the third message). However, it made the assumption that EMD had some meaning other than a mispunched END so provided three other messages which could confuse the novice programmer. Oft what makes a good programmer is the ability to interpret error messages.

```
          PROGRAM
        C DIST
        C A PROGRAM TO COMPUTE DISTANCES BETWEEN POINTS
        C READ DATA VALUES
    1           READ(5,1)X1,Y1,X2,Y2
    2        11 FCRMAT(4F5.0)
        C COMPUTE DISTANCE
    3           DIST=SQRT((X1-X2)**2+(Y1-Y2)**2)
        C PRINT ANSWER
    4           WRITE(6,2)DIST
    5         2 FORMAT('0',F7.0)
    6           STOP
    7           EMD
**WARNING**   UNNUMBERED EXECUTABLE STATEMENT FOLLOWS A TRANSFER
***ERRCR***   UNDECODEABLE STATEMENT
**WARNINC**   MISSING END STATEMENT;END STATEMENT GENERATED
**WARNINC**   END STATEMENT NOT PRECEDED BY A TRANSFER
***ERROR***   MISSING FORMAT STATEMENT       1 USED IN LINE       1
**WARNING**   FCRMAT STATEMENT       11 IS UNREFERENCED
```

Exercises and Problems

1   Which of the following variable names are incorrect?

```
IJKLMN
MOTHER
A+26
A26
88R4
XXX1234
```

2   Write the following as arithmetic expressions in FORTRAN:
$24x^2 + 19x - 38$
$$\frac{(a-b)^2}{(a+b)^2}$$
$(26 - 14y)^{-1}$

3   If P has the value 10.5, what value will replace the current value of K in the following assignment statement?

```
K=P/4.2
```

4   What would the outcome be if a FORMAT statement number were punched in column 6 of a standard punch card instead of column 5? What if the number w punched in column 7?

5   Which of the following are executable statements?

```
READ
WRITE
FORMAT
STOP
END
```

6   Rewrite DIST so that it prints the input data (x and y coordinates) as well as th distance between the two points. (This is called an echo check; many programmers incorporate such checks in all their programs.)

7   Write a FORTRAN program which computes the distance between two points given both horizontal (x and y) and vertical (z) coordinates of each point.

0906274

064714

3

FORTRAN for Spatial Data

The assignment, input and output statements which were described in the previous chapter have been defined only for single variables (also called scalars). One of the characteristics of spatial problems is that the data will almost certainly occur as a set of numbers, either as a matrix of values or as vectors of coordinates and values. The various techniques by which maps are coded into these forms are considered in chapters 4 and 5. For the moment, we will use a simple example for purposes of illustration. Assume a dot map showing the distribution of a phenomenon such as buildings, people or plants. By drawing or superimposing a grid on this map (figure 3.1), it is possible to count the number of occurrences in each cell and to form a table or matrix which represents the map in numerical form (table 3.1).

Table 3·1　A dot map in numerical form

8	4	4	4	1	1	2	0
3	5	2	6	1	2	3	1
1	3	2	8	3	4	1	1
1	1	2	4	4	3	1	1
2	3	2	5	2	2	1	1
1	2	3	5	1	2	1	2
1	1	3	3	1	2	1	1
2	3	4	1	2	3	1	2
1	1	2	2	3	1	1	2
1	2	2	2	1	1	1	0

There are 80 numbers in this table. If it were necessary to refer to each cell by a different variable name in a FORTRAN program, we would be faced with some tedious coding and punching to do any calculation. To find the mean value, for example, would require an assignment statement whose right side would have all the cell names separated by addition signs, obviously an impractical procedure even for small arrays. This kind of problem should never occur in FORTRAN (or any other computer language) because it is possible to identify any array by a single variable name, and to refer to particular elements by subscripts.

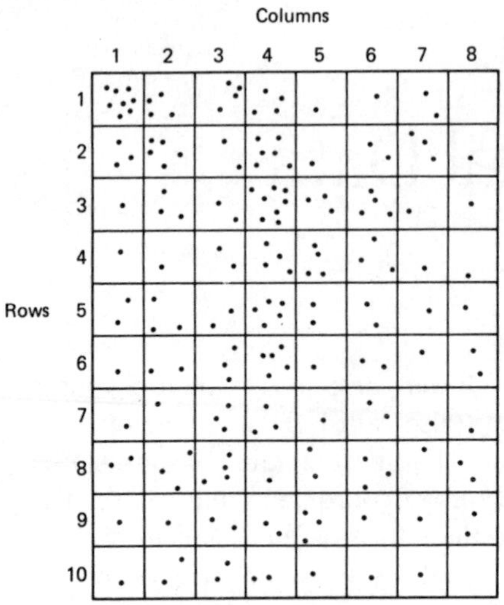

Figure 3.1   A hypothetical dot map.

For example, a vector or string of numbers such as

9. 87. 4. 103.2 67.1 2. 0.2

could be labelled with the variable name XYZ. The third element in the vector (4.) would then be referred to as XYZ(3), and the seventh (0.2) as XYZ(7). In the case of a matrix such as table 3.1, two subscripts are required to identify a particular element, and the variable name used to label the data set is said to be doubly subscripted, or to have two dimensions. The convention for these subscripts is that the first designates the row, and the second the column (where rows go across and columns go up and down). If the matrix in table 3.1 were identified by the variable name KOUNT, then the upper left element (8) would be designated as KOUNT(1, 1), and that in the second row, fourth column (6) would be labelled KOUNT(2, 4).

This chapter is primarily concerned with computing operations with subscripted variables. The statements necessary for assignment, input, and output of arrays are described and illustrated with three examples. Other FORTRAN statements are introduced and illustrated which allow conditional execution and branching within programs, and the capability for defining subprograms.

3.1   DIMENSION Statements

Before a subscripted variable may be used in a computer program, it must be identified at the outset so that the compiler can set aside a sufficient number of memory locations. This is done with a DIMENSION statement, one of a class of FORTRAN statements termed specification statements. These are not executable, but declare space requirements and special conventions at the outset of a program.

The general form of a DIMENSION statement for a matrix is

DIMENSION NAME, (m, n)

where m and n are integer constants giving the maximum number of rows and columns to be reserved for the matrix NAME. In the case of a vector, only one subscript would appear in the DIMENSION statement, for example,

```
DIMENSION PQR(10)
```

and in the case of a triply subscripted variable

```
DIMENSION VOL(4,8,6)
```

(Most FORTRAN compilers allow up to seven subscripts, but it is difficult to conceive of many problems which require such data structures.)

3.2   Arrangement of Arrays in Computer Memory

A vector is stored in the memory of a computer as a string of values in an ascending sequence of storage locations. Arrays with two or more dimensions are also stored as strings of numbers by the machine by placing each column after another in a sequence of locations.[1] For example, a matrix dimensioned as B(2, 3) would be allocated to storage in the following sequence:

B(1, 1) B(2, 1) B(1, 2) B(2, 2) B(1, 3) B(2, 3)

A three-dimensional array such as C(3, 2, 3) would be arranged in computer storage as

C(1, 1, 1) C(2, 1, 1) C(3, 1, 1) C(1, 2, 1) C(2, 2, 1) C(3, 2, 1) C(1, 1, 2) C(2, 1, 2)
C(3, 1, 2) C(1, 2, 2) C(2, 2, 2) C(3, 2, 2) C(1, 1, 3) C(2, 1, 3) C(3, 1, 3) C(1, 2, 3)
C(2, 2, 3) C(3, 2, 3)

Thus the general rule is that the first subscript of an array increases most rapidly and the last increases least rapidly.

The method by which the machine arranges its storage is normally of no importance to the computer programmer. When he specifies that a variable be used in an assignment statement, he is not interested in its specific location in the computer memory. In fact, programming languages such as FORTRAN have been written specifically to free him from such considerations. However, arrays with two or more dimensions present special problems. Programs for such data which take computer memory organization into account will execute many times faster than those which do not. This is explained further in this chapter in a discussion of a feature called virtual memory, and in the section describing input and output of arrays.

3.3   DO Statements and Loops

When subscripted variables are used in assignment, input, output and other executable statements, the operation is controlled by a DO statement. The general form of this statement is

DO n $i=j, k, m$

where n is the number of a statement following the DO statement which is the last to be controlled by it, i is an integer variable called the *DO variable*, and j, k, and m are integer variables or constants called the *initial value*, the *test value*, and the *increment* of the statement, respectively. These define how the statement or statements under the control of the DO statement are to be executed. This can be illustrated with an example:

```
     DO 23 I=1,12,1
23      A9(I)=A8(I)/ALPHA
```

---

[1] In many programming languages, including PL/I, BASIC, and APL, rows are chained together, rather than columns.

These two statements form the simplest kind of a DO loop, a single assignment statement which is repeatedly executed under the control of a DO statement. When the DO statement is first encountered in the computer program, the DO variable is set to 1, the initial value of the loop, and statement 23 is executed. The two sub-scripted variables in this assignment statement, A9 and A8, use the DO variable as a subscript. This means that A9(1) and A8(1) are used the first time this statement is executed.

Control then returns to the DO statement, and I, the DO variable, is increased by 1, the increment of the loop. Statement 23 is executed for a second time, now with the subscript of each vector set to 2. This is repeated, with I increasing by 1, until I equals the test value of the loop, in this case 12. When the DO variable reaches the test value, the loop is executed for the last time, and control passes to the statement which follows the DO loop. Thus statement 23 is executed 12 times, and 12 different elements of each of the vectors A9 and A8 are involved in the calculation.

If the increment of the loop were 2 instead of 1, that is

```
DO 23 I=1,12,2
```

then statement 23 is executed with the DO variable equal to 1, then 3, 5, 7, 9, and 11. At this point, the comparison of the DO variable with the test value of the loop indicates that if it is incremented once more, it will exceed 12, even though it is not equal to the test value at the time of the check. To accommodate such cases, the test is actually whether the DO variable equals the highest possible value which does not exceed the test value. In this example, the loop would be executed six times, and every second element of the subscripted variables would be used in the assignment statement.

The increment of a DO loop is almost always 1. For this reason, FORTRAN com-pilers allow a programmer to delete it from the DO statement. The increment is then set by default to 1. The DO statement above then becomes

```
DO 23 I=1,12
```

When dealing with two-dimensional arrays, it is necessary to use two DO state-ments to control computations for all elements. For example, we may have two gridded maps 10 rows by 8 columns which are to be summed and stored in a third:

```
      DO 12 J=1,8
      DO 12 I=1,10
12    X(I,J)=A(I,J)+B(I,J)
```

When two DO statements are involved, we refer to an inner (DO variable I) and an outer (DO variable J) loop, and state that the inner loop is nested within the outer loop. The pattern of execution of assignment statement 12 is that J is first set to its initial value 1, then the inner loop is completely executed, that is, I is set to 1 through 10, and the subscripts of the matrix elements used in the calculation are (1, 1), (2, 1), (3, 1), (4, 1),..., (10, 1). When the inner loop is satisfied, the DO variable of the outer loop is incremented by 1, and the inner loop completely executed again, that is, I is set to 1, then 2, etc., and the subscripts are (1, 2), (2, 2), (3, 2), (4, 2), (5, 2), and so on. This pattern is repeated until the DO variable of the outer loop equals the test value (8).

3.4 DO Loops and Virtual Memory

Virtual memory or virtual storage is a feature which is becoming increasingly common on intermediate and large computer systems. The principle of virtual memory is that the user of the machine has the illusion that the core memory available for a program is far larger than the actual space physically available in the machine. The IBM 370 used at the University of Pennsylvania, for example, has 3 million bytes of actual core, but 16 million bytes which are apparently available to the programmer. The illusion of a very large memory is maintained by the operating system of the computer by storing most instructions and data on a peripheral storage device (typically a disk) which has a very large capacity but a slow access time (the time required to read data from the unit into core, or to write data from core to the unit). Instructions and data which are not immediately required in a particular program are transferred to the peripheral storage unit in units called pages, segments of several thousand bytes (in IBM 370s, 4000 bytes), and moved back into core memory just before they are required by the CPU (central processing unit).

Virtual memory may present a particular problem for the programmer concerned with spatial problems because his data sets are so large that they will occupy several pages. For example, a matrix of 60 rows by 40 columns will total 2400 elements, each requiring four bytes of storage. This means that at least three pages will be required on an IBM 370. The fact that a data set spans several pages is of no real importance in a computer program provided that there are a minimum number of requests for values stored at locations in other pages. Unfortunately, it is easy in FORTRAN to arrange DO statements so that page changes occur frequently, so often, in fact, that a trivial re-arrangement in the order of DO statements can make the program run as much as several hundred times faster.

Consider the following DO loop:

```
      DO 65 I=1,60
      DO 65 J=1,40
65    ACE(I,J)=0.0
```

The first DO statement refers to the row subscript and the second the column subscript, probably the most natural way to refer to a matrix computation. In the execution of these statements, the DO variable I is set to 1, and J proceeds from 1 through 40. The matrix ACE is stored in memory as a vector, ordered columnwise from the matrix. This means that as J is incremented by 1, the element in the vector addressed in core storage is actually 60 locations further on. The loop thus progresses through core in steps 60 elements long. While I has the value 1 at least three pages will be referenced by the inner loop. When the entire loop has been completed, over 200 pages will have been referenced. If the loop is re-arranged so that the DO statements are reversed:

```
      DO 65 J=1,40
      DO 65 I=1,60
65    ACE(I,J)=0.0
```

then the elements of ACE addressed in memory are adjacent to one another because of the columnwise ordering of the matrix. This means that page transfers occur only three or four times.

A general rule is apparent for FORTRAN on virtual memory machines: that in nested loops referring to arrays with more than one subscript, *the innermost DO statement should refer to the first subscript position, and the outermost to the last subscript.*

3.5   Input and Output of Subscripted Variables

DO statements are used with READ and WRITE statements as well as with assignment statements, for example

```
      DO 89 K=1,M
      DO 89 L=1,N
89    READ(5,46) XYZ(L,K)
```

This block of coding instructs the machine to input from punched cards values for elements of the matrix XYZ according to FORMAT statement 46. The sequence of operations is that K is set to its initial value 1, L is also set to 1, and a value for the element XYZ(1, 1) is read from a data card. L is then incremented by 1, and the value for XYZ(2, 1) is read, then XYZ(3, 1), XYZ(4, 1),...XYZ(N, 1). The innermost DO loop is then satisfied, and one column of the matrix has been read into core memory. K is then incremented to 2, L initialized again to 1, and the second column of XYZ is read.

The important thing about this block of coding is that no matter how many numbers are on each card, or how many data fields may be specified in the FORMAT statement, the READ statement inputs only one number from one card when it is executed.[2] When the column subscript is incremented by 1 and the READ executed again, it skips to the next card and reads a single number. This means that one card is required for each element of a matrix being read; an array dimensioned as 40 by 60 would thus require 2400 data cards. WRITE statements under the control of DO statements operate in a similar way. No matter what kind of output format is used, only one value will be printed on each line. Each time the WRITE is executed, it skips to the following line. These are obviously undesirable constraints since it is convenient to be able to punch several numbers on each card, and it is essential to have the ability to write a series of numbers on one line across an output page.

These problems are overcome in FORTRAN with a procedure by which DO statement information is included as part of the READ or WRITE statement. For example,

```
      DO 89 K=1,M
89    READ(5,46) (XYZ(L,K),L=1,N)
```

The expression $L=1$, N is called an *implied DO*. It contains the same information as a DO statement, except that the word DO and the statement number are omitted. The sequence of operations is that K and L are both set to their initial values (1), and the READ statement executed for element XYZ(1, 1). Then L is incremented by 1 and the value for XYZ(2, 1) read, but without moving to a new card unless required to by the FORMAT statement. If, for example, the format were

```
46    FORMAT(10F5.0)
```

and there were 35 numbers in each column of the matrix (i.e. the value of N was 35),

---

[2] This is true for all subscripted variables. Matrices are used here as an example because they are so important in spatial problems.

then the first ten elements of the first column of XYZ would be read from the first card, then the input operation would skip to the second card because the FORMAT statement had in effect been used up. The eleventh through twentieth elements in the first column would then be read from the second card, then the twenty-first through thirtieth from the third card. The fourth card would have only five numbers on it, corresponding to the thirty-first through thirty-fifth elements in the first column. When the thirty-fifth value has been read then the implied DO has been satisfied. The DO variable of the explicit DO is now incremented to 2, and the implied DO starts through from 1 to 35 again. However, the READ operation does not pick up where it left off (i.e. half way across the fourth card), but commences with a new card. The general rule is as follows: *input and output operations under the control of explicit DOs will skip to a new card or line whenever the DO variable is incremented, while input and output controlled by an implied DO will continue on the same card or printer line until the FORMAT is exhausted or the DO is satisfied.*

If implied DOs are nested, for example

```
READ(5,46)  ((XYZ(L,K),L=1,N),K=1,M)
```

then the shift from card to card is completely under the control of the FORMAT. When the end of a column is reached part way along a data card, the machine expects the first element in the following column to be located in the next position on the same card.

If matrix data are ordered columnwise on cards (column 2 of the matrix following column 1, column 3 following column 2, etc.), and there are as many data values as locations specified in the DIMENSION statement, then all references to DOs may be omitted, for example

```
READ(5,46)XYZ
```

will read values for all the elements of the matrix XYZ using the FORMAT statement 46, and in the order: column 1, rows 1 through N, column 2, rows 1 through N, and so on. Similarly,

```
WRITE(6,83)XYZ
```

will print the columns of the matrix XYZ on the lines of the printer page according to FORMAT statement 83.

Input and output of arrays without DOs is not only simpler to code, but is a much faster operation for the machine. It should, therefore, be considered the preferred technique. It is not always possible, however, particularly in programs which are designed for data sets of varying sizes. MSDMAP, the following example, illustrates this.

3.6   Example 2. MSDMAP: Mean and Standard Deviation of a Gridded Map

This example illustrates the use of DIMENSION statements, assignment under the control of DO statements and input and output of subscripted variables. A new element of FORTRAN is presented which allows titles to be placed on output pages, and a programming procedure is introduced which sums subscripted variables.

The mathematics of this example is straightforward. The mean is found by summing all values of the matrix elements, and dividing the total by the number of elements.

The standard deviation may be calculated in a number of ways.[3] The most direct is to subtract the value of each element from the mean, square this value, sum them, divide by one less than the number of elements, and take the square root of this quantity. In the case of one-dimensional data, this would be

$$s = \sqrt{\{\textstyle\sum_{I=1}^{N} (X_I - \bar{X})^2\}/(N-1)}$$

where $\bar{X}$ is the mean, and there are N data values.

In the two-dimensional case, with M rows and N columns:

$$s = \sqrt{\{\textstyle\sum_{I=1}^{M} \sum_{J=1}^{N} (X_{IJ} - \bar{X})^2\}/(M*N-J)}$$

Summation is required for both the mean and standard deviation calculations. This is often a baffling procedure for the novice programmer. If a number of single variables (scalars) are to be summed, a single assignment statement is adequate, as

```
SUM=A+B+C+D+E
```

With subscripted variables, the computation must be done under the control of DO statements, and a quite different procedure is used, that of accumulating the total in a single variable. Before the summation loop is started, the variable which is to be the accumulator is initialized to zero, as in

```
C USE SUM AS AN ACCUMULATOR
      SUM=0.0
```

Then DO statements and an assignment statement do the summation:

```
      DO 20 J=1,N
      DO 20 I=1,M
   20 SUM=SUM+X(I,J)
```

Before the loop is started, SUM has the value 0. The first time the loop is executed, SUM is set to X(1, 1), the second time it is set to X(2, 1)+X(1, 1), the third time to X(3, 1)+X(2, 1)+X(1, 1). When the inner loop is satisfied, SUM will contain the total of values for the first column of X. When the outer loop has been satisfied, SUM will contain the sum of all the values for the entire matrix X.

Let us now consider the FORTRAN statements for this example. It is conventional to start with comment cards which identify the program and describe its purpose. Program identification is often with a name representing an acronym. For example, we might name this example MSDMAP (*m*ean and *s*tandard *d*eviation of a *map*):

```
C MSDMAP                                                      MSDM    1
C A PROGRAM TO COMPUTE THE MEAN AND STANDARD DEVIATION        MSDM    2
C OF A MAP STORED AS A MATRIX                                 MSDM    3
```

The first kind of FORTRAN statement which occurs in any program after the introductory comments are specification statements which state space to be allocated for arrays and special conditions which apply for the duration of the program. The only one of these described thus far is the DIMENSION statement, required to identify all variable names which are arrays and to specify the space to be reserved for them. In this problem, there is only one array, a matrix used to store the gridded map. Since we are using data from a map such as that shown in figure 2.3, we need a matrix 10 rows by 8 columns:

```
      DIMENSION X(10,8)                                       MSDM    4
```

The next step is to read the data into core memory. Let us assume that the map

---

[3] Any readers unfamiliar with the concept of standard deviation should consult a statistics reference, or study this example without attempting to understand its utility.

has been punched on cards by columns, with ten fields per card and each field F5.0. There will be a total of eight cards, one for each column.

```
C READ THE DATA                                            MSDM   5
      DO 12 J=1,8                                          MSDM   6
12    READ(5,73)(X(I,J),I=1,10)                            MSDM   7
73    FORMAT(10F5.0)                                       MSDM   8
```

The mean value of the matrix is now computed using the summation procedure described above:

```
C USE SUM AS AN ACCUMULATOR                                MSDM   9
C FIRST INITIALIZE SUM TO ZERO                             MSDM  10
      SUM=0.                                               MSDM  11
C SUM ALL VALUES                                           MSDM  12
      DO 90 J=1,8                                          MSDM  13
      DO 90 I=1,10                                         MSDM  14
90    SUM=SUM+X(I,J)                                       MSDM  15
C DIVIDE SUM BY NUMBER OF ELEMENTS IN MATRIX TO GET MEAN   MSDM  16
      AVG=SUM/80.                                          MSDM  17
```

The mean value (stored in AVG) could be saved and printed out with the standard deviation at the end of the program, but it is good practice to write out results as soon as they are computed in case there is an error later on which terminates execution without writing out any results at all. Some kind of identification with the mean value is also very useful, otherwise we simply get a number printed out at the top of the page with no clue to its significance. Placing titles or identification in the output is done by including in the FORMAT statement the desired text immediately after the carriage control character at the beginning of the FORMAT information, and within the apostrophes that lead and follow the carriage control symbol. For this problem, it might be done as follows:

```
C PRINT RESULTS WITH TITLE                                 MSDM  18
      WRITE(6,53) AVG                                      MSDM  19
53    FORMAT('0THE MEAN VALUE IS',F6.1)                    MSDM  20
```

Now the standard deviation of the map matrix is computed using AVG and the summation principle described above. Since the total of all elements of the matrix is no longer required, the variable SUM can be used again in this section.

```
C                                                          MSDM  21
C COMPUTE STANDARD DEVIATION                               MSDM  22
C FIRST RESET SUM TO ZERO                                  MSDM  23
      SUM=0.                                               MSDM  24
      DO 123 J=1,8                                         MSDM  25
      DO 123 I=1,10                                        MSDM  26
123   SUM=SUM+(AVG-X(I,J))**2                              MSDM  27
      SDEV=SQRT(SUM/79.)                                   MSDM  28
      WRITE(6,17)SDEV                                      MSDM  29
17    FORMAT('0THE STANDARD DEVIATION IS',F7.2)            MSDM  30
      STOP                                                 MSDM  31
      END                                                  MSDM  32
```

As a final note in this example, it should be pointed out that on many machines the apostrophe or single quote method for carriage control and title information does not work. Instead one must use the H field in the FORMAT statement. This is used by first counting the number of characters which are to be printed, including blanks and the carriage control character. This number is then placed at the beginning of the format information, followed by the letter H, and then the text. For example:

```
17    FORMAT(26H0THE STANDARD DEVIATION IS,F7.2)
```

3.7   Conditional Execution (The Logical IF Statement)[4]

Thus far we have considered elements of FORTRAN which allow a programmer to read data, compute with assignment statements, print results, and stop a program. A further fundamental part of all computer languages is the ability to control execution of statements depending upon the outcome of certain tests. This can be done in two ways in FORTRAN: the first of these, conditional execution, is presented in this section; the second, program transfer, is described later.

The FORTRAN statement which is used for conditional execution is the logical IF statement, which has the general form

IF (*t*) *s*

where *t* is a logical expression (to be described below), and *s* is any executable statement except DO or another IF. This means that an assignment, READ, WRITE, or STOP statement (plus others yet to be described) is actually part of the logical IF statement, and is executed only on the condition that the expression *t* is true. If *t* is false, statement *s* is bypassed, and execution passes to the next statement in the program. If *t* is true, execution also passes to the next executable statement after *s* has been executed, except when statement *s* is itself a program transfer instruction (to be discussed later).

The logical expression in the IF statement consists of constants, variables, or arithmetic expressions separated by relational operators. The six relational operators of FORTRAN are as follows:

Operator	Meaning	Mathematical symbol
.EQ.	equal to	$=$
.NE.	not equal to	$\neq$
.GT.	greater than	$>$
.GE.	greater than or equal to	$\geq$
.LT.	less than	$<$
.LE.	less than or equal to	$\leq$

Note that the relational operators are always preceded and followed by a decimal point (in order to distinguish them from variable names).

Some examples of logical IF statements are as follows:

```
IF(GAMMA.LT.ALPHA)WRITE(6,2)X
IF(ABC.EQ.675.2)X=34.2
IF(ABC(K+1).GT.ABC(K))STOP
```

In many cases, it is useful to be able to couple logical expressions within one IF statement. This is done in FORTRAN with the logical operators .AND. and .OR.[5] They are used as follows (where *a* and *b* represent logical expressions, that is, variables, constants, or arithmetic expressions separated by relational operators):

---

[4] The logical IF statement is not part of USA Standards Institute FORTRAN IV, but is available on most large computers. Any reader who is using a machine without this capability should read this section to understand the purpose of the statement, but must use program transfer techniques to solve the same kind of programming problems.

[5] A third logical operator—.NOT.—is not described in this book.

Expression	Meaning
$(a.\text{AND}.b)$	The entire expression is true (and the statement in the second part of the logical IF executed) only if both expressions a and b are true; if either are false, the entire expression is false.
$(a.\text{OR}.b)$	The entire expression is true if either expression a or b is true; it is false only if both a and b are false.

Some examples of logical IF statements using both relational and logical operators are as follows:

```
IF(A.EQ.B.AND.C.LT..5)WRITE(6,432)A,B,C
IF(R+B/4..GT.Q.OR.A.LT.HOLD)STOP
```

There is a hierarchy of operations with relational and logical operators, just as there is with arithmetic operators. Relational operators are executed before logical operators, and both take place after arithmetic expressions are evaluated. This means that parentheses are rarely required within a logical IF statement to establish a particular order of evaluation. The default hierarchy is usually the appropriate one. (But there is nothing wrong in adding redundant parentheses: in fact, it often simplifies the formulation of involved logical expressions.)

3.8   Program Transfer

The logical IF statement provides the capability to execute a single statement only when certain conditions are true. Entire blocks of program statements can be skipped, or execution returned to an earlier statement by the GO TO statement. This has the form

GO TO n

where n is the number of an executable statement. When a GO TO instruction is encountered, control passes to the numbered statement; it is executed, as are the following statements until another transfer or an IF statement is encountered. The GO TO is almost always used with a logical IF, for example,

```
IF(A.LT.B)GO TO 18
```

In FORTRAN II and with compilers with a minimal capability, there is no provision for the logical IF, rather the arithmetic IF. This has the form

IF (t) i, j, k

where t is an arithmetic variable or expression, and $i, j,$ and k are statement numbers. If the expression t has a negative value, control transfers to statement number i; if it has the value zero, transfer is to statement j; if it is positive, control goes to statement k. The arithmetic IF is thus equivalent to three logical IFs and GO TOs. The arithmetic IF is also available on compilers which provide the logical IF, but the latter is so much more suitable for most programming problems that it has virtually replaced the arithmetic IF.

3.9   Flowcharting

When programs become relatively complicated, particularly when branching operations are included through IF and GO TO statements, it is common to provide a diagram which illustrates major steps in the flow of execution. A flow chart for a computer program is organized as a series of frames or symbols connected by arrows. Each frame represents one type of computer operation, such as assignment or input/output. Figure 3.2 presents the six flowcharting symbols about which there is agreement among most authors and organizations.

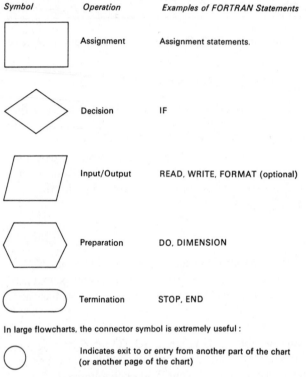

Symbol   *Operation*   *Examples of FORTRAN Statements*

Assignment   Assignment statements.

Decision   IF

Input/Output   READ, WRITE, FORMAT (optional)

Preparation   DO, DIMENSION

Termination   STOP, END

In large flowcharts, the connector symbol is extremely useful :

Indicates exit to or entry from another part of the chart
(or another page of the chart)

Figure 3.2   Principal flowcharting symbols.

A flowchart is prepared by arranging the appropriate frames in a sequence down the diagram, with branching operations placed to the right of the main axis of flow. Flowchart frames are connected by arrows (which should not cross) which indicate the path of flow. Statements under the control of a DO statement are further identified by a line on the left side of the chart from the last statement in the loop back to the DO frame. Within each frame is placed a description of the operation which takes place at that step. If the program is small and the flowchart detailed, one FORTRAN statement may be placed within each box. This leads to unwieldy and even confusing diagrams with larger programs, so that it is common to group operations within frames, and to describe them in words rather than in computer statements.

Figure 3.3, a flowchart for example 2, MSDMAP, illustrates how frames are linked together. Note particularly the DO loops and the use of the connector symbol. If the details of input and output are important in a program, format information and explicit and implied DOs are added to the input and output frames, but it is usual to

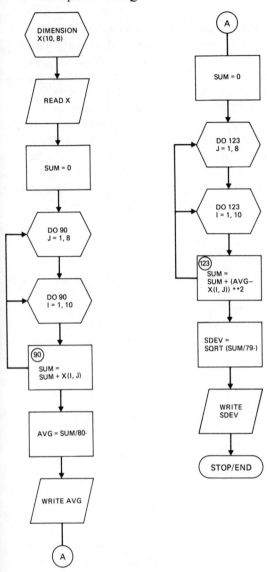

Figure 3.3   Flowchart for MSDMAP.

omit such details. The initial frame containing DIMENSION statements is also commonly deleted. A flowchart with branching (an IF statement) will be presented with the following example.

3.10 Example 3. FREQ: Frequency Distribution Tabulation

This example uses all the FORTRAN statements which have been presented up to this point,[6] demonstrates two programming techniques (finding the minimum or maximum of a set of numbers, and computing subscripts), and illustrates methods for

[6] Except GO TO and the arithmetic IF.

flexible input and attractive output. A frequency distribution is a tabular summary of
the numbers of values which fall within certain classes, and does not normally require
any mathematical operations. In this program, however, it is assumed that the user
knows only the number of classes he desires in the table. This means that the program
must examine the data to determine the class interval and the boundary values for
each class. The table is then built up through counting values in each class, an
inspection operation when done manually, but an assignment operation in a computer.

After the initial comment and DIMENSION statements, the program is divided into
four main parts. The first reads the data, the second sorts through these data to find
the maximum and minimum (and thus the class interval), the third tabulates the
frequency distribution, and the last part of the program prints out the resulting table
of numbers. A flowchart for the complete program is presented in figure 3.4.

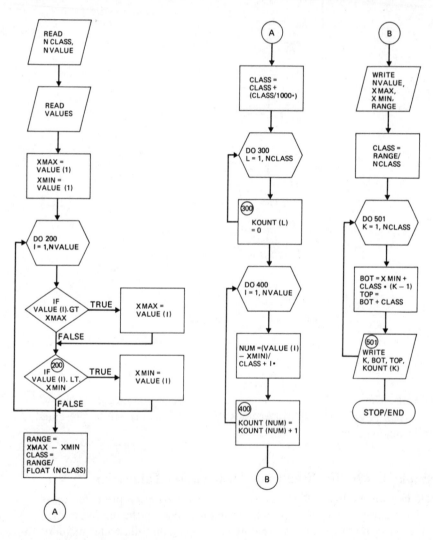

Figure 3.4   Flowchart for FREQ.

It is quite unusual that a computer program is written to be used only once. Most often a program is run with several different data sets or by several different users. For this reason, it is good programming practice to design the input section of all programs with flexible requirements. In FREQ, the user specifies his requirements as data: the first card in the data deck contains the number of classes desired in the frequency distribution table, and the total number of values in the data cards which follow. The first READ statement in the program reads these numbers, and uses them throughout the remainder of the program. This means that FREQ can be used with data sets of various sizes without any changes required in the FORTRAN statements.

The maximum array size in any FORTRAN program is that stated in the initial DIMENSION statements. In FREQ the maximum number of data values is set at 100, and the maximum number of frequency distribution classes at 10. These limits may be far too low for some problems run with this program, but there is a charge for memory locations reserved by a program in most large computers, whether these memory locations are used or not. (The charge may either be in actual costs, or in slower turn-around time.) It is good practice, therefore, to keep the upper limits down, particularly since it is such an easy matter to replace the DIMENSION statement with one specifying larger values when required.

The most rigid input requirement of FREQ is that the data must be punched in a specified format (10F5.0). (A technique to provide variable format is presented in the discussion of packaging programs in chapter 7.) The program is written for a vector of values, but can be used for arrays with two or more dimensions by treating these arrays as one-dimensional for this program. The field on the first data card which specifies the number of values for a problem will contain the product of the number of rows and columns for a matrix, and this product times the third dimension for a three-dimensional array, and so on.

The FORTRAN statements for the first part of the program are as follows:

```
C FREQ                                                        FREQ   1
C PROGRAM TO COMPUTE FREQUENCY DISTRIBUTION (UP TO 10 CLASSES) FREQ   2
C OF A VECTOR OF VALUES (UP TO 100)                           FREQ   3
C      DATA VALUES ARE STORED IN VALUE, RESULTS IN KOUNT      FREQ   4
         DIMENSION VALUE(100),KOUNT(10)                       FREQ   5
C                                                             FREQ   6
C PART ONE: INPUT DATA                                        FREQ   7
C READ NUMBER OF CLASSES (NCLASS) AND NUMBER OF VALUES (NVALUE) FREQ 8
C FROM FIRST DATA CARD                                        FREQ   9
         READ(5,100)NCLASS, NVALUE                            FREQ  10
100      FORMAT (2I5)                                         FREQ  11
C READ DATA IN 10F5.0 FORMAT                                  FREQ  12
         READ(5,101) (VALUE(K),K=1,NVALUE)                    FREQ  13
101      FORMAT(10F5.0)                                       FREQ  14
```

The second part of the program examines the data values which have been read to find the maximum and minimum. The procedure for doing this is to first define variables which will contain the maximum and minimum, in this case, XMAX and XMIN. These are both set initially to the value of the first element in VALUE:

```
C                                                             FREQ  15
C                                                             FREQ  16
C PART TWO: COMPUTE RANGE AND CLASS INTERVAL                  FREQ  17
C DETERMINE MAXIMUM AND MINIMUM VALUES IN DATA                FREQ  18
C INITIALIZE XMAX AND XMIN                                    FREQ  19
         XMAX=VALUE(1)                                        FREQ  20
         XMIN=VALUE(1)                                        FREQ  21
```

Now each element of VALUE after the first is compared in turn with XMAX and XMIN in IF statements. If a data value exceeds the current value of XMAX, then XMAX is reset to the data value, or if the data value is less than the current value of XMIN, then XMIN is reset. If neither of these conditions is true, no change is made to XMAX and XMIN, and execution proceeds to the next element in VALUE. The coding for this is:[7]

```
C  GO THROUGH DATA TO FIND MAXIMUM AND MINIMUM          FREQ  22
       DO 200 I=2,NVALUE                                FREQ  23
       IF(VALUE(I).GT.XMAX)XMAX=VALUE(I)                FRFQ  24
  200  IF(VALUE(I).LT.XMIN)XMIN=VALUE(I)                FREQ  25
```

The operation of this loop is illustrated in table 3.2 for some hypothetical data. The first time the loop is executed, the second element of VALUE is less than the current value of XMAX and less than the current value of XMIN; the value of XMIN is thus reset. The second time through the loop, XMIN is again reset, but XMAX remains at its initial value until the third time the loop is executed. This process is repeated until all elements of the vector have been examined.

Table 3.2   Finding minimum and maximum values

DO variable	Data value	XMAX value before IF	IF true?	New XMAX value	XMIN value before IF	IF true?	New XMIN value
	(XMAX and XMIN initialized to 342.1)						
2	132.9	342.1	no	——	342.1	yes	132.9
3	43.0	342.1	no	——	132.9	yes	43.0
4	368.3	342.1	yes	368.3	43.0	no	——
5	205.4	368.3	no	——	43.0	no	——
6	169.3	368.3	no	——	43.0	no	——
7	375.8	368.3	yes	375.8	43.0	no	——

The range and class interval are now calculated from the maximum and minimum

```
C  COMPUTE RANGE AND CLASS INTERVAL                     FRFQ  26
       RANGE=XMAX-XMIN                                   FREQ  27
       CLASS=RANGE/NCLASS                                FREQ  28
```

The assignment statement which computes CLASS has a mixed-mode expression on its right side in that a real variable is divided by an integer variable. Most FORTRAN compilers accept such expressions, and treat the integer as a real variable (or constant). Some compilers, however, require that the FORTRAN-supplied function FLOAT be used with the integer variable in a mixed-mode expression, for example

```
       CLASS=RANGE/FLOAT(NCLASS)
```

This function converts the quantity within parentheses to its real or floating-point equivalent. (The function for converting real to integer is IFIX.)

---

[7] Some compilers will not allow a DO loop to terminate with an IF, particularly when the third part of the IF is a GO TO. In these cases, the loop is terminated with a CONTINUE statement, described in the following chapter.

The third part of the program computes the frequency distribution table. The programming procedure for this is to compute a number from each data value which can be used as a subscript in the frequency distribution vector (KOUNT). In FREQ this is done with

```
NUM=(VALUE(I)-XMIN)/CLASS+1.
```

NUM is the number of the frequency distribution class (i.e. the subscript of KOUNT) to which this data value belongs. The operation of this statement may be illustrated with some numerical examples. Let us assume a vector of numbers with a maximum of 375.8, a minimum of 43.0, and therefore a range of 332.8. If ten classes are desired in the frequency distribution table, then the class interval is 33.28.

We will trace the operation of the statement for three values: 123.0 (selected at random), 43.0 (the minimum), and 375.8 (the maximum). The first number, 123.0, has subtracted from it the value of the minimum, giving 80.0, and this is divided by the class interval, 33.28, giving 2.4. To this is added 1.0, giving 3.4. NUM is now set to this value, but since it is an integer variable, truncation will first occur; the quantity after the decimal point, 0.4, is deleted, and NUM is set to 3. Thus 123.0 belongs to the third of the ten frequency distribution classes.

The case of the minimum, 43.0, is quite simple since it is subtracted from itself, giving zero. Zero divided by any non-zero quantity is zero, so the value of the entire expression is 1.0. NUM is thus set to 1; the minimum belongs to the first class in the frequency distribution.

The case of the maximum gives an unexpected result. Its value less the minimum gives the range, and this divided by the class interval gives 10.0 (the number of classes). When this is increased by 1.0, the expression has the value 11.0. The maximum therefore belongs to the eleventh class in the distribution, a rather undesirable result since 10 classes were specified! This apparent error occurs only with the maximum value, and is overcome by changing the value of the class interval very slightly so that when divided into the range it gives a result just under 10.0 (such as 9.999999). The expression on the right of the assignment statement thus has the value 10.999999, and NUM is set after the truncation to 10, the correct answer.

The FORTRAN coding for the third part of FREQ starts with this change to the class interval:

```
C                                                              FREQ   29
C PART THREE: TABULATE FREQUENCY DISTRIBUTION                  FREQ   30
C INCREASE CLASS INTERVAL VALUE VERY SLIGHTLY TO AVOID PROBLEM FREQ   31
C WITH MAXIMUM DATA VALUE                                      FREQ   32
        CLASS=CLASS+(CLASS/1000.)                              FREQ   33
```

The values of KOUNT are all set to zero before the table calculations commence because the frequency distribution is tabulated by adding to previous values in KOUNT. The first time through the tabulation, KOUNT will have values left by the previous program in the machine since the computer does not erase information from its memory locations after each program is completed.

```
C INITIALIZE KOUNT TO ZERO                                     FREQ   34
        DO 300 L=1,NCLASS                                      FRFQ   35
300     KOUNT(L)=0                                             FREQ   36
```

The vector of frequency distributions (KOUNT) is then determined using the procedure described above:

```
C TABULATE FREQUENCY DISTRIBUTION                              FREQ  37
      DO 400 I=1,NVALUE                                        FREQ  38
C COMPUTE CLASS NUMBER FOR EACH DATA VALUE                     FREQ  39
      NUM=(VALUE(I)-XMIN)/CLASS+1.                             FREQ  40
C INCREMENT NUMTH CLASS                                        FREQ  41
  400   KOUNT(NUM)=KOUNT(NUM)+1                                FREQ  42
```

The fourth and final part of FREQ prints out the results of the calculations. First a title is written, together with values for the number of classes, the maximum, minimum, and range. This kind of information may not be required by many users of the program, but it is sometimes very useful and easily included in the coding. It is good practice to print the major results of any computation. FORMAT statement 500 illustrates the use of continuation cards, and uses a format code not yet described, the symbol /. This is the signal for the end of an output line; on encountering this symbol, the printer control unit recognizes that a new line is to start, and looks for the carriage control character.

```
C                                                             FREQ  43
C PART FOUR: OUTPUT RESULTS                                   FREQ  44
      WRITE(6,500)NVALUE,XMAX,XMIN,RANGE                       FREQ  45
  500   FORMAT('1FREQ: FREQUENCY DISTRIBUTION PROGRAM',/,      FRFQ  46
      1 'ODATA SET HAS',I5,' VALUES',/,                        FPEQ  47
      2 'OMAXIMUM IS',F6.1,/,'OMINIMUM IS',F6.1,/,             FRFQ  48
      3 'ORANGE IS',F6.1,/,                                    FREQ  49
      4 'OCLASS NUMBER  LOWER LIMIT   UPPER LIMIT   FREQUENCY') FPEQ  50
```

Printing the frequency distribution table is tricky because the actual class boundaries have not been used in the computation. It is necessary to compute them simply for the purpose of complete output. BOT and TOP are the variables used for the lower and upper boundaries of each class. They are re-computed each time through the output loop. The coding for this section follows:

```
C RESET CLASS INTERVAL TO EXACT VALUE                         FREQ  51
      CLASS=RANGE/FLOAT(NCLASS)                                FRFQ  52
      DO 501 K=1,NCLASS                                        FREQ  53
      BOT=XMIN+CLASS*(K-1)                                     FRFQ  54
      TOP=BOT+CLASS                                            FPEQ  55
  501   WRITE(6,502)K,BOT,TOP,KOUNT(K)                         FRFQ  56
  502   FORMAT('0',6X,I2,8X,F7.1,6X,F7.1,3X,I4)                FREQ  57
      STOP                                                     FREQ  58
      END                                                      FREQ  59
```

Some typical output from FREQ is presented below:

```
FREQ: FREQUENCY DISTRIBUTION PROGRAM

DATA SET HAS    25 VALUES

MAXIMUM IS 157.0

MINIMUM IS    1.0

RANGE IS 156.0
```

CLASS NUMBER	LOWER LIMIT	UPPER LIMIT	FREQUENCY
1	1.0	32.2	5
2	32.2	63.4	12
3	63.4	94.6	3
4	94.6	125.8	2
5	125.8	157.0	3

3.11   Subroutine Subprograms

It is common that a programmer finds he has need for a particular block of coding such as DIST, MSDMAP, or FREQ not as complete programs, but as parts of other programs. For example, a program may be required which reads the coordinates of a set of points, computes the distance among them, and calculates the mean, standard deviation, and frequency distribution of these distances. It is possible to assemble such a program by merging parts of DIST, MSDMAP and FREQ, but a much simpler approach available in most computer languages is to convert each of the three original programs into a self-contained subprogram which can be invoked by a main program when that particular computation is required. FORTRAN provides for two kinds of subprograms, the function (described later in this chapter), and the subroutine.

The subroutine subprogram is written much as a standard program in that it may contain any and all of the FORTRAN statements which have been described thus far, and they are arranged in the same way as in a regular program. It differs from a program in several ways. First, a new FORTRAN statement is placed at the beginning of the subroutine to identify it, and to indicate what information is to be transferred between it and the calling program (or subprogram). This statement is of the general form

SUBROUTINE name (arguments)

where name is a label the programmer selects for the subroutine. The rules for forming this name are the same as those for variable names, that is, six characters or less starting with an alphabetic character and no special symbols. Within parentheses in the SUBROUTINE statement are the arguments, a list of variable names which represent the information shared by this and the calling program. Because of this information transfer characteristic, it is relatively unusual (but not incorrect) to have input and output statements in a subroutine. The necessary READ and WRITE statements tend to be placed in the main program.

A subroutine is almost never terminated by a STOP and END statement pair, because this would stop execution of the entire job and any results which were computed in the subroutine would not be transferred back to the calling program. Instead a RETURN and END statement pair are used. RETURN is an executable statement which indicates that control is to be returned to the calling program.

The subroutine is invoked by the calling program by another new FORTRAN statement, of the general form

CALL name (arguments)

where name is the name of the subroutine being invoked (there may be several in the program), and the arguments are a list of variable names or constants which correspond with the argument list in the SUBROUTINE statement. The arguments in the SUBROUTINE statement are often described as *dummy* arguments since the FORTRAN compiler replaces them with names from the CALL argument list (described as the *actual* arguments) during conversion of the FORTRAN to machine language instructions.

As an example, let us consider a subroutine which calculates the mean of a two-dimensional array of numbers. The main program reads the array and then calls the subroutine. It must send the name of the array and provide a variable name to receive back the calculated mean. Thus

```
DIMENSION X(10,20)
READ ......
.
.
.
CALL MEAN(X,AVG)
WRITE ......
.
.
.
STOP
END
```

The subroutine coding is placed directly after the END of the main program:

```
   SUBROUTINE MEAN(DATA,AVG)
   DIMENSION DATA(10,20)
   SUM=0.
   DO 1 J=1,20
   DO 1 I=1,10
 1 SUM=SUM+DATA(I,J)
   AVG=SUM/200.
   RETURN
   END
```

Note that the names in the argument lists of the CALL and SUBROUTINE statements need not be identical. For example, the data matrix is called X in the main program and DATA in the subroutine. The rule is that the argument lists must agree in number, type and order of arguments, but need not match in name. In fact, it is common to use constants as arguments in the CALL statement, which are then matched with variable names in the SUBROUTINE statement.

3.12   Adjustable Dimensions

The requirement that argument lists match in type, order and number means that any array name which is an argument must be dimensioned to the same limits in both the calling program and the subroutine. This rule could result in considerable inconvenience if DIMENSION statements had to be rewritten each time a subroutine is used with a different main program. This is not necessary, however, because FORTRAN provides a feature termed adjustable or object-time dimensions which allows dimension information to be transmitted from the calling program to the subroutine and then used in the DIMENSION statement of the subroutine and for any other purpose which might be useful. The example above could thus be rewritten as follows:

```
DIMENSION X(10,20)
READ ......
.
.
.
CALL MEAN(X,AVG,10,20)
WRITE ......
.
.
.
STOP
END
```

```
SUBROUTINE MEAN(DATA,AVG,NROWS,NCOLS)
DIMENSION DATA(NROWS,NCOLS)
SUM=0.
DO 1 J=1,NCOLS
DO 1 I=1,NROWS
1    SOM=SUM+DATA(I,J)
AVG=SUM/(NROWS*NCOLS)
RETURN
END
```

Adjustable dimensions may be used only in subroutines and only for subscripted variables which appear in the argument list. If an array is used only in the subroutine, it must be dimensioned with the usual integer constants. The arguments which contain the dimension information are generally used elsewhere in the subroutine, most typically to control DO loops as illustrated above. They may not be re-defined to another value in the subroutine, however, which would occur if the variable names I and J were used in the argument list and the DIMENSION statement; the DO statements would attempt to redefine I and J, an error which should terminate execution of the program.

A common programming error in using adjustable dimensions with spatial data occurs when only part of a dimensioned matrix area is actually set with values in a READ or assignment statement in a main program, then this matrix is sent to a subroutine. In the coding above, for example, perhaps only a 5 by 6 area is required for a particular data set. When the subroutine is called by

    CALL MEAN(X,AVG,5,6)

the program will be returned either with an error message or gibberish.

The reason for a problem is as follows. As described in the section on virtual memory, the matrix X in the calling program is actually stored in the machine as a vector 200 elements long. A 5 by 6 portion of this matrix will occupy 30 locations in this vector: 1 to 5, 11 to 15, 21 to 25, 31 to 35, 41 to 45 and 51 to 55. All other positions will be undefined or have those values left by the last user of the core memory area.

In the subroutine, the matrix DATA is stated as having dimensions 5 by 6; the compiler thus expects the data to be located in a vector 30 elements long, starting in the same memory location as X. This means that only the first column of DATA will have the desired values; all other locations will either be undefined, or have erroneous values.

The solution to this problem is to ensure that the number of rows in the DIMENSION statement in the calling program is the same as the number of rows sent to the subroutine as an argument and used in an adjustable dimension. Because of the way matrix data are arranged into a vector for storage in the machine, the number of columns in a DIMENSION can be larger than the number sent as an argument.

3.13   Example 4. LISTR: Printing Spatial Data

When operating with a rectangular matrix describing a spatial distribution (such as that presented in figure 2.3 and table 3.1), it is often desired either to list the data or the result of a computing operation in some form which resembles a map. At a very minimum, this means printing numbers which are spaced equally in the row and column directions on the printer page (as in the lower part of figure 3.5). This will not

```
21.  42.  63.  84. 105. 126. 147. 168. 189. 210. 231. 252. 273. 294. 315. 336. 357. 378. 399. 420. 441. 462. 483. 504. 525.
22.  44.  66.  88. 110. 132. 154. 176. 198. 220. 242. 264. 286. 308. 330. 352. 374. 396. 418. 440. 462. 484. 506. 528. 550.
23.  46.  69.  92. 115. 138. 161. 184. 207. 230. 253. 276. 299. 322. 345. 368. 391. 414. 437. 460. 483. 506. 529. 552. 575.
24.  48.  72.  96. 120. 144. 168. 192. 216. 240. 264. 288. 312. 336. 360. 384. 408. 432. 456. 480. 504. 528. 552. 576. 600.
25.  50.  75. 100. 125. 150. 175. 200. 225. 250. 275. 300. 325. 350. 375. 400. 425. 450. 475. 500. 525. 550. 575. 600. 625.
26.  52.  78. 104. 130. 156. 182. 208. 234. 260. 286. 312. 338. 364. 390. 416. 442. 468. 494. 520. 546. 572. 598. 624. 650.
27.  54.  81. 108. 135. 162. 189. 216. 243. 270. 297. 324. 351. 378. 405. 432. 459. 486. 513. 540. 567. 594. 621. 648. 675.
28.  56.  84. 112. 140. 168. 196. 224. 252. 280. 308. 336. 364. 392. 420. 448. 476. 504. 532. 560. 588. 616. 644. 672. 700.
29.  58.  87. 116. 145. 174. 203. 232. 261. 290. 319. 348. 377. 406. 435. 464. 493. 522. 551. 580. 609. 638. 667. 696. 725.
30.  60.  90. 120. 150. 180. 210. 240. 270. 300. 330. 360. 390. 420. 450. 480. 510. 540. 570. 600. 630. 660. 690. 720. 750.
31.  62.  93. 124. 155. 186. 217. 248. 279. 310. 341. 372. 403. 434. 465. 496. 527. 558. 589. 620. 651. 682. 713. 744. 775.
32.  64.  96. 128. 160. 192. 224. 256. 288. 320. 352. 384. 416. 448. 480. 512. 544. 576. 608. 640. 672. 704. 736. 768. 800.
33.  66.  99. 132. 165. 198. 231. 264. 297. 330. 363. 396. 429. 462. 495. 528. 561. 594. 627. 660. 693. 726. 759. 792. 825.
34.  68. 102. 136. 170. 204. 238. 272. 306. 340. 374. 408. 442. 476. 510. 544. 578. 612. 646. 680. 714. 748. 782. 816. 850.
35.  70. 105. 140. 175. 210. 245. 280. 315. 350. 385. 420. 455. 490. 525. 560. 595. 630. 665. 700. 735. 770. 805. 840. 875.
36.  72. 108. 144. 180. 216. 252. 288. 324. 360. 396. 432. 468. 504. 540. 576. 612. 648. 684. 720. 756. 792. 828. 864. 900.
37.  74. 111. 148. 185. 222. 259. 296. 333. 370. 407. 444. 481. 518. 555. 592. 629. 666. 703. 740. 777. 814. 851. 888. 925.
38.  76. 114. 152. 190. 228. 266. 304. 342. 380. 418. 456. 494. 532. 570. 608. 646. 684. 722. 760. 798. 836. 874. 912. 950.
39.  78. 117. 156. 195. 234. 273. 312. 351. 390. 429. 468. 507. 546. 585. 624. 663. 702. 741. 780. 819. 858. 897. 936. 975.
40.  80. 120. 160. 200. 240. 280. 320. 360. 400. 440. 480. 520. 560. 600. 640. 680. 720. 760. 800. 840. 880. 920. 960.1000.
41.  82. 123. 164. 205. 246. 287. 328. 369. 410. 451. 492. 533. 574. 615. 656. 697. 738. 779. 820. 861. 902. 943. 984.1025.
```

Figure 3.5 Equally spaced numbers with a six line per inch printer.

happen fortuitously, because the geometry of character printing is biased; on computer printers there are typically ten characters per inch along lines (across the page), and either six or eight lines per inch down the page. This means that if it were possible to represent each element of a matrix by a single integer, then a printout of this matrix would give an exaggeration of 5/3 in the vertical direction for a six line per inch machine, and somewhat less (5/4) with an eight line per inch machine.

In order to obtain spatial fidelity on a computer printer, it is necessary to represent each grid cell by a matrix of line and character positions. In the case of a six line printer, a matrix of three lines and five characters will be exactly half an inch a side, the smallest combination which will give equal dimensions on all sides. On an eight line printer, the matrix must be four lines by five characters. (Six line printers will be assumed in this book, since they are more common.)

If the map matrix is sufficiently small, it is a simple matter to print it in a square format. For example, a matrix X dimensioned to 20 rows and 25 columns may be printed on a six line machine with the following statements.[8]

```
      DO 15 I=1,20
15    WRITE(6,830)(X(I,J),J=1,25)
830   FORMAT('-',25F5.0)
```

[8] These statements are, of course, highly inefficient for a virtual memory machine. In the coding for the example later, an assumption will be made which produces a more efficient operation.

The carriage control character of the FORMAT statement ('−') indicates triple spacing and thus gives the desired spacing in the vertical direction. The field specifications of the FORMAT statement forces output to the desired arrangement along each line.

Two problems arise with this simple approach. First the format may not be suitable for the particular data values involved. With F5.0 fields, any elements with values less than 0.5 will be printed as 0., and those greater than 9999. or less than —999. will be printed as asterisks (because there are too many digits). There are three ways to allow for situations like this:

1   use a different field specification such as F5.1, F5.2, etc. to accommodate digits after the decimal (assuming this does not cause asterisks for other values in the data);
2   use a larger matrix of character locations for one grid cell such as six lines by ten characters (discussed at the end of this section);
3   scale the data so that it fits the desired field specification.

The third alternative may appear to be most suitable in that it keeps the map at a reasonable size and does not require changing FORTRAN statements for each data set. However, it results in a listing of numbers which are considerably changed from the original data so that they fit into the available space. It is often difficult to make sense of them. In this example, it will be assumed that the data fit into F5.0 fields.

The second problem with the procedure described above is that it can be used only with data sets below a certain size. If the matrix X were 30 columns, for example, then the WRITE statement above would be

```
15    WRITE(6,830)(X(I,J),J=1,30)
```

but a FORMAT statement such as

```
830    FORMAT(30F5.0)
```

is impossible because it specifies more fields than will fit on a normal printer page (132 characters). However, if a smaller FORMAT such as

```
830    FORMAT(26F5.0)
```

is used, the first 26 elements of X will be printed on one line, the next four on a second line directly underneath it, then the first 26 elements of the next column will be printed on the following line. This output bears little resemblance to a map-type display.

This problem of data sets which are too wide is handled by printing the matrix in segments. Assuming an example with 52 columns and 30 rows, and 35 fields per line on a page (125 characters), then the following coding would print the elements with regular spacing:

```
      DO 11 I=1,30
11    WRITE(6,12)(X(I,J),J=1,25)
12    FORMAT('-',25F5.0)
C SKIP TO TOP OF NEW PAGE BEFORE NEW MAP SECTION
      WRITE(6,13)
13    FORMAT('1')
      DO 20 I=1,30
20    WRITE(6,12)(X(I,J),J=26,50)
      WRITE(6,13)
      DO 25 I=1,30
25    WRITE(6,12)(X(I,J),J=51,52)
```

This coding can be generalized for matrices of various sizes by making the initial and test values of the DOs (both explicit and implied) variables rather than constants. The number of rows in the data matrix is examined to see how many map sections (page widths) are required, and a DO loop is set up around WRITE statements which are executed as many times as there are map sections.

LISTR incorporates the generalized coding to print matrices of real numbers in a spatial format. It assumes a six line per inch printer and a maximum line length of 125 characters. The FORMAT field is F5.0, so the data are assumed to lie between —999. and 9999., and to have no significant digits after the decimal.

The calling program for this subroutine inputs the data from cards and calls LISTR with the appropriate arguments. Let us assume, for example, that a map of occurrences of plants of a particular species has been digitized by superimposing a transparent grid and counting the totals within each cell. The source map was 36 inches in the vertical dimension and 24 inches horizontally; the grid cell size was 0.25 by 0.25 inches, so the matrix of grid cells resulting is 144 rows by 96 cloumns.

These data may be punched on cards for input to the computer either by rows (groups of 96 elements) or by columns (groups of 144 elements). Most people prefer to organize data from a grid by rows. In FORTRAN programs, however, it is more efficient to treat the data as organized by columns because they may be read more quickly (using a READ without DOs), and can be both read and manipulated far more quickly in machines with virtual memory.

It is possible to satisfy both these demands by accepting spatial data organized on cards by rows, but considering these data in the machine to be organized by columns. Programs thus operate on the *transpose* of the matrix (figure 3.6). The original orientation of the data is restored on the output page by printing each column in the machine as a line. This is the procedure adopted in this and following examples in this book.

Let us assume in this case that the user decides to enter his data on cards by rows, using the entire card width and fields four columns wide. This means that for each row of 96 elements, there will be four cards with 20 numbers and a fifth with 16. There would be a total of 720 cards in the data deck.

Since the program considers each row of the original data to be a column, the card reading can be done without explicit or implied DO statements:

```
      DIMENSION DATA(96,144)
      READ(5,100)DATA
100   FORMAT(20F4.0,/,20F4.0,/,20F4.0,/,20F4.0,/,16F4.0)
```

The main program now transfers control to the subroutine, transmitting in the argument list the name of the matrix to be mapped and its dimensions:

```
      CALL LISTR(DATA,96,144)
```

After LISTR is completed, control is transferred back to the main program to the statement following the CALL statement, in this case:

```
      STOP
      END
```

The cards on which the subroutine have been punched are placed immediately after the END card of the main program, starting with

```
      SUBROUTINE LISTR(X,NROWS,NCOLS)                          LSTR   1
```

Note that the variable names in this argument list do not match those in the

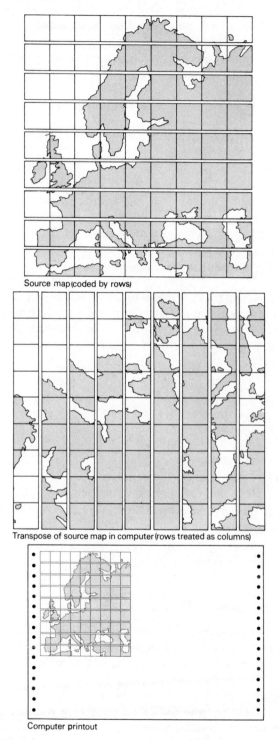

Source map (coded by rows)

Transpose of source map in computer (rows treated as columns)

Computer printout

Figure 3.6   An illustration of how a map is transposed when read into a computer, and restored to its original orientation on output.

argument list of the CALL statement in name, but do match in type, order and number.

Comment cards then describe the purpose and assumptions of the subroutine:

```
C A SUBROUTINE TO PRINT SPATIAL DATA WITH EQUAL SPACING IN          LSTR   2
C THE ROW AND COLUMN DIRECTIONS                                     LSTR   3
C USES SIX LINE PER INCH PRINTER                                    LSTR   4
C ASSUMES NO PLACES AFTER DECIMAL, AND NO VALUES > 9999. OR < -999. LSTR   5
C NO RESTRICTIONS ON SIZE OF INPUT MATRIX X                         LSTR   6
```

The matrix is now specified in the subroutine using the adjustable dimension feature:

```
      DIMENSION X(NROWS,NCOLS)                                      LSTR   7
```

Before the loops which control the output procedure are started, it is necessary to determine how many page widths are involved, and to initialize two variables which are used as location indicators. NSECS, the number of page widths must be computed so that the result is rounded upwards, rather than rounded off. This is easily done with some manœuvering of integer arithmetic:

```
C COMPUTE NUMBER OF MAP SECTIONS, ASSUMING 25 NUMBERS PER PAGE      LSTR   8
      NSECS=(NROWS+24)/25                                           LSTR   9
```

Two integer variables, KA and KB, are used to keep track of those rows in the data matrix which begin and end each map section. Their values are determined in two assignment statements at the beginning of the loop. An IF statement checks for the right side of the map by comparing KB and NROWS, and resetting it if the last map section is less than 25 elements wide.

```
C START THE OUTPUT LOOP                                             LSTR  10
      DO 13 K=1,NSECS                                               LSTR  11
      KB=K*25                                                       LSTR  12
      KA=KB-24                                                      LSTR  13
C CHECK FOR RIGHT SIDE OF MAP                                       LSTR  14
      IF(KB.GT.NROWS) KB=NROWS                                      LSTR  15
C PRINT A TITLE AT THE TOP OF EACH MAP SECTION                      LSTR  16
      WRITE(6,12)K,NSECS                                            LSTR  17
12    FORMAT('1THIS IS MAP SECTION',I3,' OF',I3)                    LSTR  18
      DO 13 J=1,NCOLS                                               LSTR  19
13    WRITE(6,11)(X(I,J),I=KA,KB)                                   LSTR  20
11    FORMAT('-',25F5.0)                                            LSTR  21
```

When all map sections have been printed the subroutine returns control to the calling program and ends the subroutine:

```
      RETURN                                                        LSTR  22
      END                                                           LSTR  23
```

3.14   LISTR with Exponential Format

If the data which need to be listed do not fit within the F5 format field, it is possible to use the E format code, a specification which will accommodate numbers of any magnitude. A number printed in E format appears as .16E 03 which translates to 0.16×10^3, or 160.

The programmer has control only over the number of significant digits when using E format. The printer reserves one space for a sign, one for the decimal, and four for the E exponent. It is obviously impossible to specify an E format which will fit five character locations. Thus the three by five character matrix used in LISTR is unsuitable, but a multiple such as six lines by ten characters will allow four significant digits, and nine by fifteen will accommodate nine significant digits.

LISTR may be modified for E format with little difficulty. The values of NSECS, KA and KB would be computed differently, and FORMAT statement 11 would be (assuming 12 fields on each line):

```
11     FORMAT('-',/,'-',12E10.4)
```

The / symbol signals the end of a line in the FORMAT. This means that the second triple space symbol '−' will be recognized as carriage control. Without the / the machine will triple space, then print the character − at the beginning of each line. The first integer following the E is the total width of the field (including spaces required by the printer), and the integer after the decimal is the number of significant digits.

The principal disadvantages of a modified LISTR are that:

1 It is difficult for the uninitiated to work with numbers in E format;
2 It produces very large quantities of paper. A 100 by 100 matrix, for example, will require almost 70 square feet of output.

3.15 Function Subprograms

Subroutine subprograms are blocks of coding which are referenced by a CALL statement in the main program or in another subroutine. Function subprograms are also blocks of coding, but they are referenced differently and generally serve a somewhat different purpose in a computer program. A basic distinction is that a function returns (and must return) a single value to the calling program. A subroutine, on the other hand, may return any number of values in the argument list, and need not return any.

Function subprograms may be called from a program, subroutine or other function subprogram by using the name of the function with its argument or arguments on the right side of an assignment statement. This has already been illustrated with the SQRT function in example 1, DIST. FORTRAN compilers provide many more such functions for operations which occur very commonly in programming problems. The following is a list of those functions which are most commonly available:

ABS( )	absolute value of a real argument
IABS( )	absolute value of an integer argument
AMAX1(, ,)*	choosing largest value from real arguments
MAX0(, ,)*	choosing largest value from integer arguments
AMIN1(, ,)*	choosing smallest value from real arguments
MIN0(, ,)*	choosing smallest value from integer arguments
FLOAT( )	conversion of an integer argument to real form
IFIX( )	conversion of a real argument to integer form
ALOG10( )	common logarithm (base 10)
ALOG( )	natural logarithm (base e)
SIN( )	self-explanatory
COS( )	self-explanatory
TAN( )	self-explanatory
COTAN( )	self-explanatory
ARSIN( )	self-explanatory

* any number of arguments may be used; three positions are shown here for convenience **only**.

ARCOS ( )   self-explanatory

SQRT( )   self-explanatory

A function subprogram written by a programmer is a block of coding which is included in the FORTRAN deck, but separate from the main program (as are subroutines). The initial statement is

FUNCTION name (arguments)

The name selected for the function must conform to the rules for forming variable names, that is, it must be six characters or less, contain no special symbols, and be of the correct type, integer or real. The name type must correspond with the value of the function *returned* to the calling program. Thus a function which returns a real value must have a name beginning with any alphabetic character other than *I, J, K, L, M* or *N*. (This is the reason for the unusual names of some FORTRAN-supplied functions such as those for logarithms and maximum and minimum values; the A is prefixed to the natural name simply to satisfy the rules for naming functions.)

A function subprogram is written in the same way as a subroutine in that a RETURN statement is required instead of a STOP, and input and output statements are unusual. The major difference is that a function subprogram must have an assignment statement which has the name of the function on the *left* side. This statement actually sets the value which is to be returned to the calling program.

3.16   Example 5. ARCDIS: Distance Between Points Given Their Latitudes and Longitudes

ARCDIS is a function subprogram which calculates the distance between two points on the surface of the earth given their latitudes and longitudes. If we assume that the earth is a sphere, then this is a problem in spherical trigonometry which involves the solution of a terrestrial spherical triangle for which the sides and the included angle are known.[9] In figure 3.7, ABC is the triangle, A is the north or south pole, B and C are two points with known coordinates (latitude and longitude), arcs AB and AC (or *c* and *b*) are their co-latitudes (90° minus their latitudes), and angle *A* is the difference in longitude of the two points. The distance between the two points is arc BC (or *a*).

The solution to this problem is given by the law of cosines:

$$\cos a = \cos b \cos c + \sin b \sin c \cos A$$

or

$$a = \cos^{-1} (\cos b \cos c + \sin b \sin c \cos A)$$

Since the earth is not a perfect sphere, this equation will yield results which are in error by a fraction of a per cent (although as high as two per cent in certain situations).

FORTRAN-supplied trigonometric functions can be used for this calculation, except that they require radians rather than degrees as arguments. Since latitudes and longitudes are almost always provided in degrees, the ARCDIS function must do a conversion first. When this is completed, a single assignment statement will compute the distance.

[9] For those with some knowledge of spherical trigonometry, this is obvious. Those unfamiliar with the subject are advised to proceed on faith.

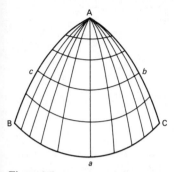

Figure 3.7   A spherical triangle.

The initial statements in the subprogram declare the function name and arguments, and describe with comment cards the purpose of the function:

```
      FUNCTION ARCDIS(LAT1,LONG1,LAT2,LONG2)                     ARCD   1
C FUNCTION TO COMPUTE DISTANCE BETWEEN TWO POINTS GIVEN          ARCD   2
C THEIR LATITUDES AND LONGITUDES, AND ASSUMING THE EARTH         ARCD   3
C IS A SPHERE                                                    ARCD   4
```

Natural names have been used for latitudes and longitudes, but since these names all start with the letter L they represent integer rather than real variables. This convention can be overridden by a new FORTRAN statement, the REAL specification. In ARCDIS, this is done with

```
      REAL LAT1,LONG1,LAT2,LONG2                                 ARCD   5
```

These particular variable names are real whenever they are used in this function subprogram.

Four assignment statements now set values for the constant pi, the co-latitudes of the two points, and the longitudinal difference between the two points. The angles are converted to radians at the same time:

```
      PI=3.14159                                                 ARCD   6
C COMPUTE ARC DIFFERENCES IN RADIANS                             ARCD   7
      AB=((90.-LAT1)*PI)/180.                                    ARCD   8
      AC=((90.-LAT2)*PI)/180.                                    ARCD   9
      BC=ABS(((LONG1-LONG2)*PI)/180.)                            ARCD  10
```

The final assignment statement sets the value of the function which is to be returned to the calling program. (The constant is the radius of the earth in miles.) Note that this statement has the name of the function on its left side:

```
C COMPUTE ARC DISTANCE                                           ARCD  11
      ARCDIS=ARCOS(COS(AB)*COS(AC)+SIN(AB)*SIN(AC)*COS(BC))*3969.665  ARCD  12
```

The subprogram is then terminated and control returned to the calling program:

```
      RETURN                                                     ARCD  13
      END                                                        ARCD  14
```

Below is presented an example of a main program which might be used with ARCDIS. The statement which actually calls the function subprogram is the assignment statement between statements 1 and 10. The remainder of the program reads coordinates off data cards and prints out results. Note that the first READ statement reads from the first data card the number of pairs of points in the data deck, and uses this value (*M*) as the test value in a DO loop which contains the call to ARCDIS. This means that the same main program can be used for data decks of various sizes.

```
      C MAIN PROGRAM FOR ARCDIS
      C READ NUMBER OF DISTANCE COMPUTATIONS TO BE MADE
    1            READ(5,99)M
    2    99      FORMAT(I5)
    3            DO 10 I=1,M
    4            READ(5,1)A,B,C,D
    5    1       FORMAT(4F7.2)
    6            DIST=ARCDIS(A,B,C,D)
    7    10      WRITE(6,2)A,B,C,D,DIST
    8    2       FORMAT('0THE DISTANCE BETWEEN',F7.2,',',F7.2,' AND',F7.2,',',F7.2,
             X ' IS', F10.1,' MILES.')
    9            STOP
   10            END
```

The cards on which ARCDIS have been punched are placed immediately after the main program, then the required system card(s), and the data. Some sample output is presented below; the results from the fourth calculation indicate the accuracy of the constants used.

```
THE DISTANCE BETWEEN   12.47,   23.35 AND   79.59,   45.00 IS   4703.8 MILES.

THE DISTANCE BETWEEN   89.10,  147.53 AND   15.48,  -97.10 IS   5189.8 MILES.

THE DISTANCE BETWEEN  -90.00,    0.00 AND   90.00,    0.00 IS  12469.7 MILES.

THE DISTANCE BETWEEN    0.00, -180.00 AND    0.00,  180.00 IS      2.4 MILES.

THE DISTANCE BETWEEN   45.00,   78.30 AND  -90.00,   78.20 IS   9353.3 MILES.
```

Problems

(Some suggestions about solutions follow).

1   Write FORTRAN statements which will read a series of five maps coded as matrices, each 12 rows by 16 columns, into a three dimensional array. (Make any assumptions you wish about how the data are punched on cards.)

2   Write FORTRAN statements which will convert a matrix of numbers into a vector by stringing column after column. (This operation is termed concatenation)

3   Two matrices were derived by placing the same grid over two maps of the same area. When transferring these data to cards, however, one matrix was arranged by rows and the other by columns. Write FORTRAN statements which will read these two matrices so that they occupy arrays with the same dimensions in the machine.

4   Rewrite example 1, DIST, so that it can be used with any number of pairs of points.

5   Rewrite example 2, MSDMAP, so that it may be used with any matrix of real numbers up to 100 by 100.

6   Rewrite FREQ as a subroutine.

7   Add coding to LISTR so that it prints column numbers across the top and line numbers down the right margin of the matrix area.

8   Example 5, ARCDIS, requires that its arguments be in degrees. Write a main program which reads coordinates in degrees, minutes, and seconds, and converts these to degrees before calling ARCDIS.

9   Write a program which finds the location and value of the maximum and minimum in a matrix, and computes the horizontal distance between them (in grid cell units). Assume no ties.

10   Write a program which reads the x and y coordinates of five points, computes a five by five matrix of the distances between all the points, and determines for each point the number of the other point which is nearest (the nearest neighbour).

11   An elevation map was digitized as a 25 by 25 matrix. There was a lake on the map, however, whose exact elevation was unknown when the map was coded. As a result, all cells which were occupied by the lake were given a zero elevation. Write a program which will compute the mean elevation of the land area, the percentage of the map occupied by lake, and (assuming the elevation of the lake is 529.4 feet) the mean elevation of the entire map.

Suggestions for the Solution of Problems

1   The usual practice is that the first two subscripts of a three-dimensional array refer to rows and columns, and the third the level or map number. The DO variable referring to the third subscript should be defined in the outermost DO statement.

2   The vector must have been specified in a DIMENSION statement to a size as large as the product of the row and column dimensions of the matrix. The concatenation consists of two assignment statements within a loop controlled by two DO statements. The first of these assignment statements computes the subscript of the vector corresponding to the row and column position, and the second assigns an element from the matrix to its appropriate position. The subscript vector can be determined by the expression
$$(J-1)*M+I$$
where I and J are DO variables referring to rows and columns, and M is the total number of rows. It may also be determined by setting a variable to 0 before the loop, and adding 1 to its previous value each time the loop is executed.

4   This may be done in several ways. The simplest is to read from the first data card the test value of a DO statement, which is used to control the number of times the distance computation its previous value each time the loop is executed. illustrated in examples 13 and 14, chapter 6.

5   The actual number of rows and columns are read from the first data card, and used as the test values of loops which follow.

7   The column numbers across the top are computed in a vector with a DO loop, then printed with an appropriate format. The row numbers on the right margin are the value of the DO variable for that line.

8   A different variable is used to store each part of the angle. The variables containing minutes and seconds are each converted to their degree equivalents and added to the variable containing degrees.

9   This will require six IF statements, two to save the trial maximum and minimum, and four to save their coordinates. This coding is quite inefficient, however, because the IF statements must be executed even when it is known that a matrix location is less than the current trial maximum and greater than the current trial minimum. A more efficient procedure is to test a value against the current trial maximum or minimum only once, and if the value does not qualify, go to the

next location in the matrix. This is most easily done with the CONTINUE statement, which is described in the following chapter. It could be used for this problem in the following way:

```
      IF(X(I,J).LT.XMAX) GO TO 25
      XMAX=X(I,J)
      ROW=I
      COL=J
25    CONTINUE
```

4

Regularly Spaced Data with Implicit Coordinates

One of the most useful methods for coding and storing spatial data is that used in MSDMAP and LISTR in chapter 3—a rectangular matrix of values derived from a grid placed on the source map. This form of data organization is well suited to computer mapping with a line printer. LISTR is an illustration of how such data may be printed in numerical form. In this chapter, five subroutines are presented which substitute characters for matrix values, printing and overprinting these to achieve the appearance of different texture patterns or densities of grey. All are written so that they may be run on either a standard or virtual memory system. As with LISTR, it is assumed that the map has been coded by rows, but the machine reads each row as a column, thus storing the transpose of the source map. On output, all the examples presented here print each column of the matrix in the machine as a line on the printer page, thus restoring the original orientation of the map.

The principal disadvantage of the matrix form for spatial data is that a very large number of data values must be collected to fill the entire matrix. Chapter 5 discusses techniques in which many fewer data values are required because they are spaced irregularly over the map (such as at boundaries, on peaks, or in the centre of homogeneous areas), with a possible cost in reduced fidelity of the computer representation of the source map. Even with these techniques, however, the data must be converted to matrix form if they are to be mapped with a line printer.

Data stored in the matrix form are said to have implicit coordinates because location information is not specified for each element but is implicit in the method of storage. This means that distance calculations require that the coordinates of each element be explicitly calculated using its row number, column number, and cell size. For example, the distance between the centre of a cell at row I, column J, and one at row K, column L would be given by

```
DIST=SQRT(FLOAT(I-K)**2-FLOAT(J-L)**2)*CELL
```

where CELL is the length of the side of a cell. If spatial data have explicit coordinates, then distance calculations are done with coding such as that presented in example 1, DIST.

4.1 Map Coding into Matrix Form by Manual Techniques

The process of coding a map so that it may be read into a computer (also termed geocoding or map digitizing) may be done in several ways, depending on the type of map, the precision required, and the expected use of the data. In this and the follow-ing section, techniques are discussed in which the map is coded as a matrix. It is assumed that no machines are available to assist in the task.[1]

Maps which are dot distributions are converted into matrix form by the procedure described in chapter 3 for figure 3.1. A grid is drawn on the map or on a transparency which is then superimposed on the map, and points are counted within each cell.

If the data source is an isarithmic map, the same general procedure of applying a grid to the map applies, but the values coded are those which occur at specific points —the intersection of the grid lines (termed grid, mesh, lattice, pivotal, or nodal points).[2] These values are estimated by visual interpolation from adjacent isarithms, a simple and relatively precise technique. (The author has found that most people can work to one-tenth of the contour interval with no difficulty.) Figure 4.1 is a hypo-thetical map which has been digitized into a matrix form using mesh points; note how the grid has been offset relative to the boundaries of the map.

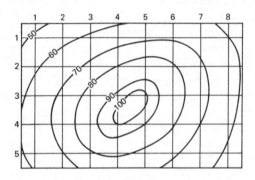

		Column						
	1	2	3	4	5	6	7	8
1	51	56	61	66	69	69	66	60
2	59	66	73	80	83	81	71	63
Row 3	63	72	83	94	101	85	73	63
4	64	74	86	100	90	81	70	62
5	63	70	79	81	78	72	65	59

Figure 4.1   Isarithmic map as interpolated values at intersections of grid lines.

Choropleth maps may be coded as matrices in three general ways, using points, lines, and areas:

1   The technique just described for isarithmic maps may also be applied to choropleth maps: a grid is superimposed on the source map and those regions which fall exactly at the intersection of grid lines are coded (figure 4.2). (If the data values are to be analysed with statistical procedures, or there are strong periodicities in the distributions, the points should be arranged in a random fashion as described in chapter 5.)

2   Lines which traverse the map along the centrelines of rows or columns may be used together with a grid to code a choropleth map; the type or value over which

[1] Readers interested in a detailed discussion of map encoding procedures should consult Tomlinson (1972).
[2] These are all terms commonly used in connection with the numerical solution of partial differential equations, a topic with many concepts useful in spatial analysis.

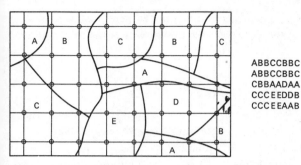

ABBCCBBC
ABBCCBBC
CBBAADAA
CCCEEDDB
CCCEEAAB

Figure 4.2   Coding at intersections of grid lines.

most of the line passes within a cell is coded as the type or value in that cell
(figure 4.3). (A more efficient use of traverse lines is described in the following
chapter.)

3   The most common technique is to use the grid cell itself, and to code that type or
value which occupies *most* of the cell, that is, the modal type or value (figure 4.4).

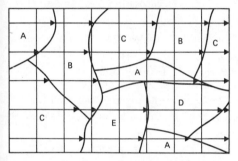

AABCCBBC
ABBCCBBC
CBBAAAAC
CCCEEDDD
CCCEEAAB

Figure 4.3   Coding by length of traverse line.

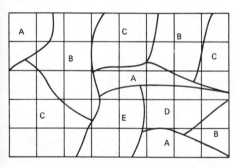

AABCCBBC
ABBCCBBC
CBBAAADC
CCCEEDDB
CCCEEAAB

Figure 4.4   Coding by modal type in grid cell.

If the original map consists of qualitative data, the resulting matrix of values will
comprise the names of regions or types. Names or character data are read in
FORTRAN in the same way as numerical data except for a different format field
(described in a later section). Once in the machine, however, maps coded as name
labels may pose substantial problems because of the time required for common
operations. As a result, it is usual to convert name labels to number labels in the

machine (described in MAP1 later). This problem is mentioned here because it is sometimes easier to make the conversion from names to numbers at the map coding stage, rather than later.

4.2   The Grid Size

Once a map encoding method has been chosen, it is then necessary to decide on the spacing of lines in the grid which is drawn or superimposed on the source map. Several considerations are important in this decision.

First, the grid must be fine enough to capture the level of detail important to the problem for which the data are being collected. If the problem is simply to transfer a map into the computer in a form which has reasonable fidelity to the original,[3] then it is possible to make some general rules about spacing in the grid. Mesh points on choropleth maps, for example, should be spaced less than the shortest distance across the smallest region of interest on the map, measured along both the row and column axes of orientation of the grid. In this way, at least one point will fall within the smallest region on the map. The rule for lines on choropleth maps is the same as for points except that the measurement for shortest distances across regions is made at right angles to the intended direction of the traverse lines. If grid cells are used to digitize a choropleth map, the cell size must be small enough to ensure that the smallest region of interest on the map is the modal type in at least one cell. This means that the grid spacing must be no more than half the distance across the smallest region of interest (and thus twice as dense as the grid required for points). On isarithmic maps, the points must be spaced no more than the distance between the two most closely spaced isarithms (the steepest slope), measured along the two axes of the grid.

These general rules define the most suitable grid spacing as that required to capture the details on a particular map. If this map is then revised by the addition of more detail, it is quite possible that a different grid spacing would be required. These guides would also result in a different grid spacing for a series of maps, even if they were of the same area. Clearly it is often necessary to select a grid using criteria other than the cartographic detail of a particular map.

Many maps are encoded to be compared with other maps of the same area, either as part of a spatial information system to identify concurrences of given conditions, or to test hypotheses concerning the relationship of spatial variables. In such problems all the spatial data sets must be comparable in some way, and the easiest way to achieve this is to code them using matching grids. The criteria for selecting this common grid depend on the specific problem, but include such considerations as the best grid for the most important map, the grid appropriate for the least detailed map (the equivalent of the least common denominator), and matching errors resulting from a grid spacing too close to the accuracy limits of a map.

The last of these considerations is perhaps most overlooked in map coding decisions, and perhaps most serious in its consequences. Consider the following

---

[3] 'Reasonable fidelity' is taken to mean that the essential spatial information content of the map is represented in the computer, but not the details that would allow the map to be reconstructed cartographically after it had been processed by the computer.

example. The hypothesis is advanced that forest in a certain area is significantly related to topographic elevation, soils, and rock formation. The test of this hypothesis involves digitizing maps of forest, elevation, soils, and geology (lithology), and comparing occurrences with some statistical procedures. If we assume 0.1 by 0.1 inch cells on a scale of 1:24000, and reasonably good source maps, there is a significant chance that the hypothesis will be rejected simply because of error in the data. This is because the accuracy of the match is the product of the accuracy standards of each component map. Thus if each map were 90 per cent accurate for this cell size and scale (that is, there is a 90 per cent chance that what is mapped as being within a 0.1 inch square is actually there when checked on the ground), then the probability that a given combination of forest, soil, slope, and geology exists in any one cell is only 66 per cent (0.9^4). If the accuracy of each component map is only 80 per cent, then the probability of the combination existing is only 41 per cent! Accuracies of 80 to 90 per cent are not at all unusual for cells 0.1 inch square on a 1:24 000 map, particularly near the edge of regions (such as the edge of a forest) and in complicated patterns (particularly in soils maps).[4] If data have been collected at a much smaller scale, such as 1:250 000, and re-mapped at 1:25 000, then 80 per cent accuracy would be excellent for a 0.1 inch square cell; 50 or 60 per cent would be more likely. (With scale differences such as this, significant error could also be introduced because of map projection differences, a problem not considered in this book.)

There are two ways to correct unacceptable matching errors. The first is to collect new data, verifying the combination of characteristics by field survey or aerial photography. The second and more usual way is to increase the cell size. This increases the probability that a specific value or type actually occurs close to a mesh point or within a grid cell, but decreases the accuracy of the estimate of the map values. There is a trade-off between positioning accuracy and value accuracy.

Complex choropleth maps digitized by the area method can pose particular problems if the grid size is increased. First, the digitizer is required to make much more difficult decisions about the contents of a grid cell. Determining the dominant type is quite simple if no more than three types ever occur within a cell, but if four, five, or even six types occur, the decision becomes extraordinarily complicated. At the same time, the values coded are less comparable with data collected on a mesh point or traverse line basis because of the increased area involved. In a complex choropleth map, it is quite possible that the modal type within each cell commonly covers less than 50 per cent of its area. It may be necessary in such cases to code not only the modal type, but an estimate of the proportion of the cell it occupies, or perhaps to identify secondary and tertiary types if they occupy a significant area.

The decision on grid size is clearly both important and difficult. Complicating it is the fact that a small change in cell size can make a substantial difference in digitizing time, and map digitizing is one of the more tedious exercises known to man!

---

[4] Unfortunately, map accuracy standards are difficult to determine for many maps. The US National Map Accuracy Standards, for example, state that no more than ten per cent of well defined points such as road intersections and corners of large buildings can be in error by more than 1/30 inch on maps of scales 1:20000 or larger, and that on contour maps of all scales, not more than ten per cent of elevations can be in error by more than one half the contour interval. (Revised Exhibit A to Bureau of the Budget Circular No. A-16, 1947.) Other countries use statistical measures such as standard deviation.

4.3   Character Data

A capability for manipulating character or alphameric in addition to integer and real data is essential for spatial applications to accommodate choropleth maps coded as name labels and to permit use of the standard line printer for mapping spatial data with characters to give the impression either of different textures or grey levels. Characters are read and printed in FORTRAN using the A format field. For example, the statements

```
        READ(5,90)WORD
90      FORMAT(A4)
```

would read the characters in the first four columns of the data card, and store them as the value of the variable name WORD. Similarly,

```
        WRITE(6,72)JUMP
72      FORMAT(' ',A3)
```

will print out the three characters stored as the value of JUMP.

In contrast to integer and real variable names, there is no particular convention which identifies a variable name as one used to store character information. Character variable names can start with any letter of the alphabet. However, it is important to be consistent about this if character data are to be compared in a program because the same characters are stored differently in the computer word depending on whether the variable name is integer or real. Thus the statement

```
        IF(WORD.EQ.JUMP) ....
```

will not be true if the same characters were read from data cards for both variables.

The maximum number of characters which can be stored in a computer word varies with the machine. In IBM 360 and 370 computers, four characters is the maximum for the standard computer word, and eight characters for double-length words (described in the following section). If an input format specifies more than the maximum, for example (assuming INFO is a standard length word),

```
        READ(5,87) INFO
87      FORMAT(A7)
```

only the rightmost four characters will be transmitted and stored as the value of INFO. Those characters punched in columns one to three of the data card will be ignored. If the format specifies a length less than the maximum allowed, the machine stores the characters read in the leftmost spaces, and fills in the right spaces with blanks. If the format for output is less than the maximum number of character spaces allowed, the leftmost characters in the word are used.

4.4   Data Initialization

Computer mapping with a line printer involves printing and overprinting characters in a manner corresponding with data values. The characters which are to be used for this could be read from data cards using A format, but it would be more desirable if they could be set in the program, thus avoiding the problem of just where the character data cards go in relation to other data cards. One would expect that the easiest way to do this would be with a series of assignment statements at the outset of the program. In FORTRAN, however, this is impossible; there is no way that a character constant can appear on the right side of an assignment statement. An alternative technique is available for setting variables to character values within the the program using data initialization.

The initialization feature of FORTRAN allows variables to be set to initial real, integer, or character values at the outset of a program unit (main program, sub-routine, or function). There are two ways to do this. The first (and more widely available) technique is with a DATA statement. This has the general form

DATA var/x/

where var is one or more variable names, and x is one or more real or integer con-stants, or (for character data), Hollerith literals.

Some examples of DATA statements are:

```
DATA IJK/4/
DATA CANUCK /543.987/
DATA ALPHA/4HABCD/
DATA BETA/'END'/
```

The last two examples illustrate the two ways of representing a Hollerith literal. One may either precede the characters by an H and the exact number of characters in the value (thus 4HABCD, 3HABC, or 2HAB), or simply bracket the characters with apostrophes. The latter method is so much more convenient (particularly for titles on output), that it has virtually replaced the H designation on compilers which offer both.

More than one value or one kind of value may be introduced in one DATA statement. Thus

```
DATA NAME,GAMMA,FIRST,LAST/'WE',4.65,6.98,999/
```

Values for dimensioned variables may also be introduced, providing there are at least as many values within the slashes as there are dimensioned locations for each variable. For example,

```
DIMENSION VALU(5)
DATA VALU/1.2,1.4,1.5,5.0,9.25/
```

Repeated values may be indicated by an integer which gives the number of repetitions, followed by an asterisk, then the value to be repeated. Thus

```
DIMENSION VALU(5)
DATA VALU/3*1.5,2*4.0/
```

will store 1.5 in the first three locations of the vector VALU, and 4.0 in the fourth and fifth locations. This repetition capability is particularly useful when an array must be initialized. For example,

```
DIMENSION MAP(100,100)
DATA MAP/10000*0/
```

The second and less commonly available technique for data initialization combines information from the DATA statement with a type specification statement. Type statements are declarations at the outset of a program or subprogram which override the predefined conventions for variable names. One of these already presented is the REAL statement, used with example 5, ARCDIS. Another commonly used is INTEGER. The explicit type specification may be used not only to override name conventions, but also to change word lengths provide dimension information, and to set initial values. For example,

```
REAL*8 X(2,4) /8*'XXXXXXXX'/
```

states that each element of the two-dimensional array X is real (which would have been true anyway), eight bytes in length (twice the standard word[5]), and set to an initial value of XXXXXXXX. The initialization part of the type statement is the same as the third part of the DATA statement: the two slashes and all the information between them. The length specification is the only alternative to standard (four byte) length for real variables. Integer variables may be defined as half-length (two byte) by

```
INTEGER*2 ....
```

but integer variables may not be eight bytes long.

The advantage of data or explicit type specifications for character information is clear—it is the only technique for introducing character data into a program without reading from cards. For numerical data, these statements are useful because they may reduce the number of assignment statements required (because a set of values can be set in one data statement). They also result in a reduction in storage space since an assignment statement such as

```
PI=3.14159
```

requires space for PI, the constant, and the machine language instruction for the assignment operation. The statement

```
DATA PI/3.14159/
```

or

```
REAL PI/3.14159/
```

results in the location PI being set to the constant at the compilation step when the FORTRAN is being converted to machine language instructions. No additional storage space is required.

Since initialization takes place during compilation, it is possible during execution to re-define with assignment statements variables which have been initialized, but it is impossible to re-initialize a variable with another data or explicit type specification. This means that it is sometimes necessary to initialize with assignment statements, as will be illustrated in MAP1.

4.5   Example 6. MAP1: Line Printer Maps of Qualitative Data

This and the following example describe basic procedures for mapping spatial data with a line printer. In both programs, the data are in the form of a matrix in which each cell represents a rectangle rather than a square on the original map. It is assumed that a six line printer is being used, so the ratio of sides of each rectangle (vertical:

---

[5] Compilers which do not have the length specification feature using the asterisk and the number of bytes allow for double length words by the specification DOUBLE PRECISION. This has exactly the same meaning as REAL* 8.

horizontal) is 5:3. (Other examples in this chapter describe how matrices with square grid cells are mapped with a line printer.)

Qualitative or nominal data are choropleth maps in which each region is identified by a label or name. A simple and direct procedure for handling such data sets is to code and read them as characters (the sequence on the left of figure 4.5). A map of this data set could be obtained by simply writing out the first character of the value in each cell. For example,

```
        WRITE(6,7) DATA
7       FORMAT(' ',95A1)
```

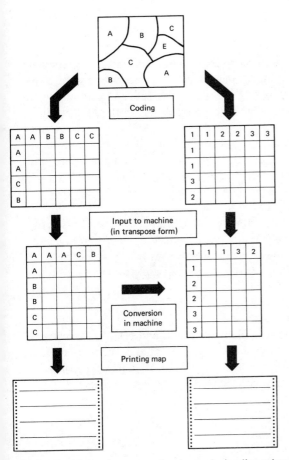

Figure 4.5   Alternative approaches to producing line printer maps of qualitative data.

Note that the use of the A1 field means that the first character of the contents of each element is printed; other characters are ignored. The principal problems with this approach are that:

1   different values which start with the same character will be mapped as the same;
2   only a small number of values can be accommodated before duplication of the first character is bound to occur.

A simple remedy for these problems is to overprint the second character in each name over the first. Duplication is much less likely, and many different values can be accommodated. (This is suggested as a programming problem at the end of this chapter.)

An alternative and somewhat more general procedure is illustrated on the right of figure 4.5. Here the source map consists of nominal values, but these have been converted to numerical labels either by the person doing the digitizing before they are read into the machine, or by a computer program which converts nominal data to numerical labels. In the map printing operation, characters are substituted for numbers with a set of statements such as

```
      DO 80 J=1,N
      DO 80 I=1,M
80    X(I,J)=SYMBOL(X(I,J))
```

where X is the data matrix and SYMBOL is a vector containing the characters which are to be used for mapping. In this loop, each element of X is replaced by the X_{IJ}th character in SYMBOL. If the number stored at a particular row and column location in X were 11, for example, then it would be replaced by the eleventh character in SYMBOL.

MAP1 is based on the procedure on the right of figure 4.5. The input data to the subroutine must be a matrix with less than 131 rows (to fit on one page width), and consisting of integers with values greater than or equal to 1, and less than or equal to 41. The reason for these limits on values is that the particular character set used consists of 40 characters plus blank.

The numerical values required by MAP1 have either been determined during the map coding, or are assigned to name values (character data) by the program before MAP1 is called. The coding for this replacement operation is relatively straightforward. For example, the names of the regions on the choropleth map and the corresponding numerical labels could be read off cards:

```
      READ(5,26) (NAME(K),NUM(K),K=1,NTYPES)
26    FORMAT(A4,I3)
```

where NAME is a vector for the names of the map values, and NUM is a vector of the numbers to be assigned. Each card would have one name and one number.

The replacement of names with numbers is done with an IF statement which compares an element of the data matrix with the vector of names. When a match is found, the name is replaced by a number:

```
      DO 100 J=1,NCOLS
      DO 100 I=1,NROWS
      DO 200 K=1,NTYPES
      IF(MAP(I,J).EQ.NAME(K)) GO TO 100
200   CONTINUE
100   MAP(I,J)=NUM(K)
```

This block of coding introduces a new FORTRAN statement, CONTINUE. This is a one word statement which is always numbered and is always referenced elsewhere in the program by either a DO statement or a GO TO. Its purpose is only to simplify the structure of a program; it literally is a 'do-nothing' statement. In this coding the CONTINUE allows control to be transferred out of the inner loop once a match is found. If instead all three DO statements terminated with

```
200   IF(MAP(I,J).EQ.NAME(K)) MAP(I,J)=NUM(K)
```

then the entire list of names would be checked at each map location, even after a
match was found.

Another approach to the replacement problem is to store the names in one matrix
and the numbers in another. For example:

```
        DO 100  J=1,NCOLS
        DO 100  I=1,NROWS
        DO 200  K=1,NTYPES
        IF(MAP(I,J).EQ.NAME(K)) GO TO 100
200     CONTINUE
100     MATRIX(I,J)=NUM(K)
```

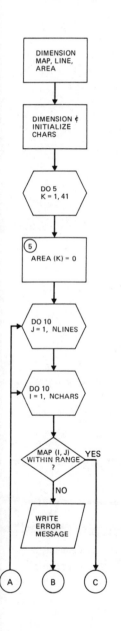

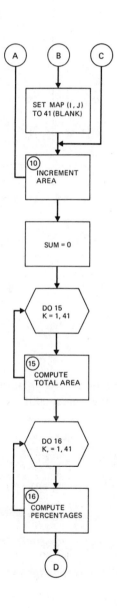

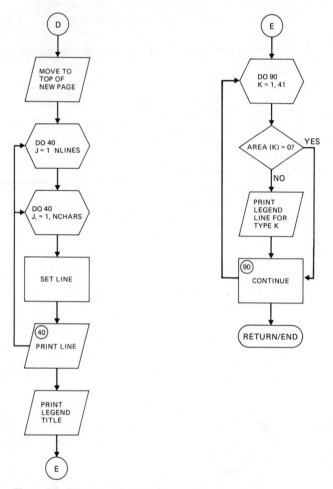

Figure 4.6   Flowchart for MAP1.

This has the advantage of keeping MAP intact if other operations must be done with it, but has the disadvantage that much additional memory space is required.

The FORTRAN coding for MAP1 starts with a set of comments and declarations:

```
      SUBROUTINE MAP1(MAP,NCHARS,NLINES)                               MAP1   1
C PRINTS MAP WITH VARYING TEXTURE PATTERNS                             MAP1   2
C INPUT DATA (MAP) IS A MATRIX OF INTEGERS > 0 AND < 41 (41=BLANK)     MAP1   3
C WHICH IS THE TRANSPOSE OF THE LINE PRINTER MAP TO BE PRODUCED        MAP1   4
C WIDTH OF MAP(NCHARS).LE.131 (MAXIMUM CHARACTERS IN A LINE)           MAP1   5
C NO RESTRICTION ON LENGTH OF MAP (NLINES)                             MAP1   6
      INTEGER MAP(NCHARS,NLINES)                                       MAP1   7
C LINE IS USED TO STORE PRINT CHARACTERS FOR ONE LINE OF OUTPUT        MAP1   8
      REAL LINE(131)                                                   MAP1   9
C AREA CONTAINS AREA OF EACH TYPE                                      MAP1  10
      DIMENSION AREA(41)                                               MAP1  11
```

The characters selected for the mapping operation are initialized in the vector CHARS by an explicit specification statement combined with data initialization:

```
C CHARS CONTAINS PRINT CHARACTERS                                      MAP1  12
      REAL CHARS(41)/'A','B','C','D','E','F','G','H','I','J','K','L',   MAP1  13
     $ 'M','N','O','P','Q','R','S','T','U','V','W','X','Y','Z',         MAP1  14
     $ '1','2','3','4','5','6','7','8','9','+','-','=','*',':',' '/      MAP1  15
```

The first set of computations determines the area of each category and checks for values outside the allowable range. The first step is to initialize the elements of the vector in which area totals will be summed:

```
C INITIALIZE AREA                                                    MAP1  16
      DO 5 K=1,41                                                    MAP1  17
5     AREA(K)=0.                                                     MAP1  18
```

Here is a case where it is important to initialize with an assignment statement rather than by a data statement or an explicit type specification. If loop 5 were replaced by

```
DATA AREA/41*0./
```

the elements of the vector would all be set to zero during compilation of the FORTRAN into machine language instructions. The subroutine would operate properly if only called once. If called a second time, however, the elements of AREA would have the value they had the last time they were used on the left side of an assignment statement.

The data checking and area calculations are straightforward:

```
C CHECK VALUES AND COMPUTE AREAS                                     MAP1  19
      DO 10 J=1,NLINES                                               MAP1  20
      DO 10 I=1,NCHARS                                               MAP1  21
C IF CK, SKIP TO END OF LOOP TO COMPUTE AREA                         MAP1  22
      IF(MAP(I,J).GT.0.AND.MAP(I,J).LE.41)GO TO 10                   MAP1  23
      WRITE(6,20) J,I                                                MAP1  24
20    FORMAT('0ELEMENT AT LINE',I3,' CHARACTER',I3,                  MAP1  25
     * ' IS OUTSIDE ALLOWABLE RANGE. IT HAS BEEN RESET TO 41 (BLANK)')  MAP1  26
      MAP(I,J)=41                                                    MAP1  27
10    AREA(MAP(I,J))=AREA(MAP(I,J))+1                                MAP1  28
C CALCULATE PERCENTAGES                                              MAP1  29
      SUM=0.                                                         MAP1  30
      DO 15 K=1,41                                                   MAP1  31
15    SUM=SUM+AREA(K)                                                MAP1  32
      DO 16 K=1,41                                                   MAP1  33
16    AREA(K)=(AREA(K)/SUM)*100.                                     MAP1  34
```

In the output section, the vector LINE is used for printing each line on the page. It is set in statement 50 to that character in CHARS which corresponds to the integer value of MAP at that location:

```
C POSITION PRINTER AT TOP OF NEW PAGE                               MAP1  35
      WRITE(6,30)                                                    MAP1  36
30    FORMAT('1')                                                    MAP1  37
      DO 40 J=1,NLINES                                               MAP1  38
      DO 50 I=1,NCHARS                                               MAP1  39
C FILL LINE WITH PRINT CHARACTERS                                    MAP1  40
50    LINE(I)=CHARS(MAP(I,J))                                        MAP1  41
C PRINT LINE                                                         MAP1  42
40    WRITE(6,60)(LINE(I),I=1,NCHARS)                                MAP1  43
60    FORMAT(' ',131A1)                                              MAP1  44
```

The legend presents for each map symbol the corresponding type number and the percentage of the map occupied. An IF statement checks the area of each type before printing in order to delete those with zero area.

```
C                                                                    MAP1  45
C MAP LEGEND                                                         MAP1  46
      WRITE(6,80)                                                    MAP1  47
80    FORMAT('1SYMBOL  TYPE NUMBER   AREA(%)')                       MAP1  48
      DO 90 K=1,41                                                   MAP1  49
C IF A TYPE HAS ZERO AREA, DELETE IT FROM THE LIST                   MAP1  50
      IF(AREA(K).EQ.0.) GO TO 90                                     MAP1  51
      WRITE(6,100)CHARS(K),K,AREA(K)                                 MAP1  52
100   FORMAT('0',2X,A1,9X,I3,9X,F5.1)                                MAP1  53
90    CONTINUE                                                       MAP1  54
      RETURN                                                         MAP1  55
      END                                                            MAP1  56
```

Figure 4.7 illustrates the main limitation of MAP1. It is difficult to separate regions from one another because the characters all tend to look the same. Problem 2 at the end of this chapter describes a modification to overcome this. (Problems 1 and 4 also suggest some useful additions.)

Figure 4.7 Some output from MAP1.

4.6   Example 7. MAP2: Line Printer Maps of Quantitative Data

MAP2 is intended for matrices of real numbers in which each element represents the number of events or occurrences within a rectangular cell (dot maps), the estimated value at an intersection on a rectangular grid (isarithmic maps), or the value of a region to which a rectangular cell belongs (choropleth maps). A line printer map of these kinds of spatial data presents each character location with a density or greyness which corresponds in some way with the numerical value of each matrix element.

The line printer with a standard character set is an imperfect instrument for obtaining a good series of grey level steps. The maximum number of levels attainable appears to be about 20 (Tomlinson, 1970), and this requires several overprints. This is a serious constraint for continuous tone patterns such as photographs. For maps, however, 20 grey levels are far more than would be useful for most applications. This is because the typical map reader is unable to distinguish more than about seven shades of grey when he is required to identify the value at a particular point or within a specified region on a map (Robinson and Sale, 1969). Seven grey levels can be easily obtained by overprinting, and the texture patterns of the symbols provide an additional aid to separating levels when viewed from a closer distance.

Table 4.1 presents some character combinations selected by the author which will give seven grey levels using one print and two overprint operations. In MAP2, characters from this matrix (called TABLE) are assigned to an output area (called LINE) with a procedure which is closely related to frequency distribution calculations (as presented in example 3, FREQ). This is because the principle of MAP2 is that the range of values is divided into equal intervals, and all elements in each class are given the same grey level.

Table 4.1   Character Combinations for Seven Grey Levels

	Grey Level						
	1	2	3	4	5	6	7
Print	.	+	=	X	+	S	M
First Overprint					/	*	E
Second Overprint							W

The FORTRAN coding begins with comments and declarations:

```
      SUBROUTINE MAP2(X,NCHARS,NLINES)                              MAP2   1
C PRINTS MAP OF SEVEN GRAY LEVELS BY DIVIDING RANGE INTO EQUAL INTERVALSMAP2   2
C INPUT DATA (X) IS MATRIX OF REAL NUMBERS NCHARS BY NLINES WHICH IS     MAP2   3
C THE TRANSPOSE OF THE LINE PRINTER MAP TO BE PRODUCED                   MAP2   4
C WIDTH OF MAP(NCHARS).LE.131 (ONE PAGE WIDTH ON STANDARD PRINTER)       MAP2   5
C NO RESTRICTION ON LENGTH OF MAP (NLINES)                               MAP2   6
C TABLE CONTAINS CHARACTERS                                              MAP2   7
C LINE IS USED TO STORE PRINT AND OVERPRINT CHARACTERS FOR ONE LINE OF   MAP2   8
C OUTPUT                                                                 MAP2   9
      DIMENSION X(NCHARS,NLINES)                                         MAP2  10
      REAL LINE(131,3)                                                   MAP2  11
```

TABLE is a character matrix which is declared and set to initial values in an explicit specification statement. The constants in the initialization section of this statement are arranged as a vector which corresponds with the columns of the matrix.

Thus the first seven elements are the values for the first column, the next seven elements those for the second, and the last seven the values for the third column. (For arrays with two or more dimensions, the general rule is that the first or inner dimension varies most rapidly.)

```
      REAL TABLE(7,3)/'.','+','=','X','+','S','M',          MAP2  12
     *               ' ',' ',' ',' ','/','*','E',           MAP2  13
     *               ' ',' ',' ',' ',' ','W'/               MAP2  14
```

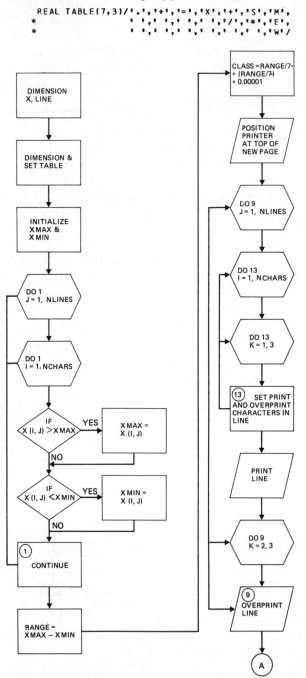

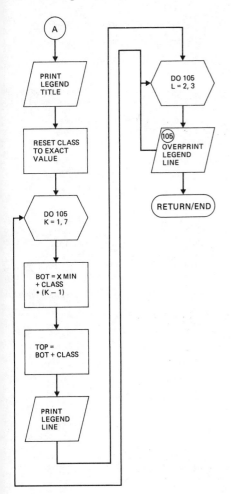

Figure 4.8   Flowchart for MAP2.

The variables determined in the initial computations in this subroutine are the same as those in FREQ: the maximum and minimum values of the array, the difference between them (the range), and the interval for one class:

```
C DETERMINE MAXIMUM AND MINIMUM VALUES IN MAP          MAP2   15
C SET INITIAL VALUES                                   MAP2   16
      XMAX=X(1,1)                                       MAP2   17
      XMIN=X(1,1)                                       MAP2   18
      DO 1 J=1,NLINES                                  MAP2   19
      DO 1 I=1,NCHARS                                  MAP2   20
      IF(X(I,J).GT.XMAX) XMAX=X(I,J)                    MAP2   21
      IF(X(I,J).LT.XMIN) XMIN=X(I,J)                    MAP2   22
1     CONTINUE                                          MAP2   23
C DETERMINE RANGE                                       MAP2   24
      RANGE=XMAX-XMIN                                   MAP2   25
C SET CLASS INTERVAL TO VALUE JUST SLIGHTLY LARGER THAN MAP2   26
C RANGE/7. TO AVOID PROBLEMS WITH MAXIMUM              MAP2   27
      CLASS=RANGE/7.+(RANGE/7.)*0.001                   MAP2   28
```

LINE is now filled with print and overprint characters from TABLE according to values in corresponding elements of the data matrix X. Note that the expression used to determine the grey level number (card MAP2 35) is the same as the arithmetic expression used to determine the class of a value in the frequency distribution program. As LINE is completed for each line, it is printed with three WRITE operations, then re-used for the following line.

```
C POSITION PRINTER AT TOP OF NEW PAGE                          MAP2  29
      WRITE(6,79)                                              MAP2  30
79    FORMAT('1')                                             MAP2  31
C ASSIGN CHARACTERS TO LINE                                    MAP2  32
      DO 9 J=1,NLINES                                         MAP2  33
      DO 13 I=1,NCHARS                                        MAP2  34
      KL=(X(I,J)-XMIN)/CLASS+1.                               MAP2  35
      DO 13 K=1,3                                             MAP2  36
13    LINE(I,K)=TABLE(KL,K)                                   MAP2  37
C PRINT                                                        MAP2  38
      WRITE(6,31) (LINE(I,1),I=1,NCHARS)                      MAP2  39
31    FORMAT(' ',131A1)                                       MAP2  40
C OVERPRINT                                                    MAP2  41
      DO 9 K=2,3                                              MAP2  42
9     WRITE(6,53)(LINE(I,K),I=1,NCHARS)                       MAP2  43
53    FORMAT('+',131A1)                                       MAP2  44
```

The legend is then printed with a block of coding which is almost identical to that used for the output of results in FREQ:

```
C LEGEND STARTS ON NEW PAGE WITH TITLE                         MAP2  45
      WRITE(6,100)                                             MAP2  46
100   FORMAT('1CLASS NUMBER  GRAY LEVEL   LOWER BOUND   UPPER BOUND')  MAP2  47
C RESET CLASS INTERVAL TO EXACT VALUE                          MAP2  48
      CLASS=RANGE/7.                                          MAP2  49
C GO THROUGH LOOP SEVEN TIMES, ONCE FCR EACH CLASS             MAP2  50
      DO 105 K=1,7                                            MAP2  51
C SET CLASS LIMITS                                             MAP2  52
      BOT=XMIN+CLASS*(K-1)                                    MAP2  53
      TCP=BOT+CLASS                                           MAP2  54
C PRINT                                                        MAP2  55
      WRITE(6,110) K,TABLE(K,1),BCT,TCP                       MAP2  56
110   FORMAT('0',5X,I2,9X,A1,7X,F11.2,2X,F11.2)               MAP2  57
C OVERPRINT                                                    MAP2  58
      DC 105 L=2,3                                            MAP2  59
105   WRITE(6,115) TABLE(K,L)                                 MAP2  60
115   FORMAT('+',16X,A1)                                      MAP2  61
      RETURN                                                   MAP2  62
      END                                                      MAP2  63
```

4.7   Example 8. MAP3 and MAP4: Line Printer Maps with Character Blocks

MAP1 and MAP2 are suitable only for data sets in which each grid cell corresponds to a rectangular area on the source map. The sides of this rectangle have the ratio 5:3 or 5:4 depending on whether a six line per inch or an eight line per inch printer is to be used. In general, coding maps with rectangular cells is not a good practice. The resulting matrix is easily mapped with a line printer, but is less appropriate than one based on square cells for other output devices such as line plotters or cathode ray tubes, and introduces significant bias to further mathematical and statistical operations on the spatial data set.

The next four subroutines in this chapter are concerned with mapping matrices coded with square grid cells with a line printer. MAP3 (for qualitative data) and MAP4 (for quantitative data) represent each matrix element on the printer page with a block of characters which is square; SQUEZ1 and SQUEZ2 modify a data matrix composed of square cells so that it can be mapped with single characters using MAP1 or MAP2.

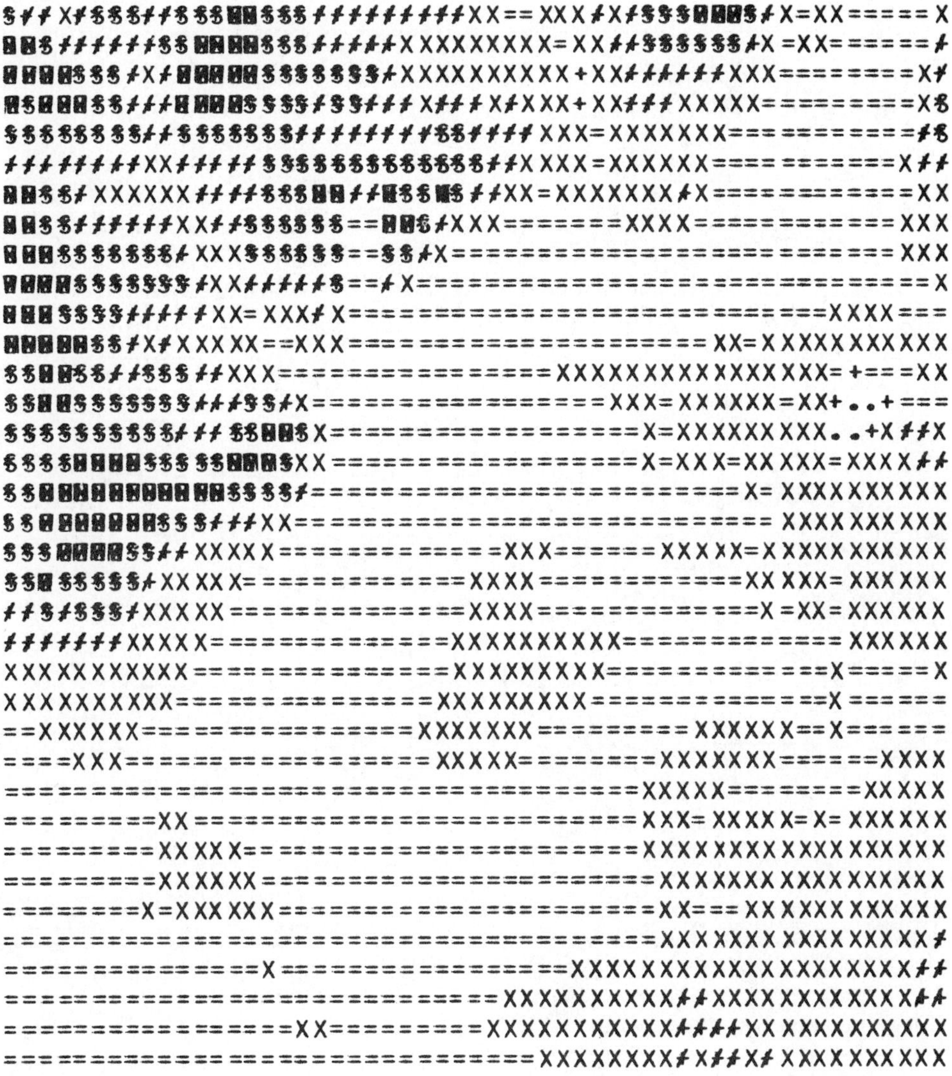

Figure 4.9   Some output from MAP2.

MAP3 and MAP4 combine MAP1 and MAP2 with the ideas presented in example 4. LISTR. Each matrix element is printed as a block three lines by five characters,[6] and the subroutines are structured to accommodate map matrices of any size. If a page width is exceeded, the subroutine prints that section of the map on a new series of pages. An example of output from MAP3 is presented in figure 4.10 and that from MAP4 in figure 4.12.

Most of the FORTRAN statements comprising MAP1 and MAP2 can be used unchanged in MAP3 and MAP4. The changes required are in the initial specifications and in the central part of each program which prints the map. In the first part of both

[6] A six line printer is assumed; an eight line per inch machine would require blocks four lines by five characters.

```
THIS IS MAP SECTION  1 OF  1
BBBBBCCCCCDDDDDEEFEEFFFFFGGGGGHHHHHIIIIIJJJJJKKKKKLLLLLMMMMMNNNNNOOOOOPPPPPQQQQQRRRRRSSSSSTTTTTUUUUUVVVV
BBBBBCCCCCDDDDDEEFEEFFFFFGGGGGHHHHHIIIIIJJJJJKKKKKLLLLLMMMMMNNNVNOOOOOPPPPPQQQQQRRRRRSSSSSTTTTTUUUUUVVVV
BBBBBCCCCCDDDDDEEFEEFFFFFGGGGGHHHHHIIIIIJJJJJKKKKKLLLLLMMMMMNNNNNOOOOOPPPPPQQQQQRRRRRSSSSSTTTTTUUUUUVVVV
CCCCCDDDDDEEFEEFFFFFGGGGGHHHHHIIIIIJJJJJKKKKKLLLLLMMMMMNNNNNOOOOOPPPPPQQQQQRRRRRSSSSSTTTTTUUUUUVVVVVWWW
CCCCCDDDDDEEFEEFFFFFGGGGGHHHHHIIIIIJJJJJKKKKKLLLLLLMMMMMNNNNNOOOOOPPPPPQQQQQRRRRRSSSSSTTTTTUUUUUVVVVVWWW
CCCCCDDDDDEEFEEFFFFFGGGGGHHHHHIIIIIJJJJJKKKKKLLLLLLMMMMMNNNNNOOOOOPPPPPQQQQQRRRPRSSSSSTTTTTUUUUUVVVVVWWW
DDDDDEEEEEFFFFFGGGGGHHHHHIIIIIJJJJJKKKKKLLLLLMMMMMNNNNNOOOOOPPPPPQQQQQRRRRRSSSSSTTTTTUUUUUVVVVVWWWHHXXX
DDDDDEEEEEFFFFFGGGGGHHHHHIIIIIJJJJJKKKKKLLLLLMMMMMNNNNNOOOOOPPPPPQQQQQRRRRRSSSSSTTTTTUUUUUVVVVVWWWHHXXX
DDDDDEEEEEFFFFFGGGGGHHHHHIIIIIJJJJJKKKKKLLLLLMMMMMNNNNNOOOOOPPPPPQQQQQRRRRRSSSSSTTTTTUUUUUVVVVVWWWHHXXX
EEEEEFFFFFGGGGGHHHHHIIIIIJJJJJKKKKKLLLLLMMMMMNNNNNOOOOOPPPPPQQQQQRRRRRSSSSSTTTTTUUUUUVVVVWWWHHHXXXXXYYY
EEEEEFFFFFGGGGGHHHHHIIIIIJJJJJKKKKKLLLLLMMMMMNNNNNOOOOOPPPPPQQQQQRRRRRSSSSSTTTTTUUUUUVVVVWWWHHHXXXXXYYY
EEEEEFFFFFGGGGGHHHHHIIIIIJJJJJKKKKKLLLLLMMMMMNNNNNOOOOOPPPPPQQQQQRRRRRSSSSSTTTTTUUUUUVVVVWWWHHHXXXXXYYY
FFFFFGGGGGHHHHHIIIIIJJJJJKKKKKLLLLLMMMMMNNNNNOOOOOPPPPPQQQQQRRRRRSSSSSTTTTTUUUUUVVVVWWWHHXXXXXYYYYYZZZ
FFFFFGGGGGHHHHHIIIIIJJJJJKKKKKLLLLLMMMMMNNNNNOOOOOPPPPPQQQQQRRRRRSSSSSTTTTTUUUUUVVVVWWWHHXXXXXYYYYYZZZ
FFFFFGGGGGHHHHHIIIIIJJJJJKKKKKLLLLLMMMMMNNNNNOOOOOPPPPPQQQQQRRRRRSSSSSTTTTTUUUUUVVVVWWWHHXXXXXYYYYYZZZ
GGGGGHHHHHIIIIIJJJJJKKKKKLLLLLMMMMMNNNNNOOOOOPPPPPQQQQQRRRRRSSSSSTTTTTUUUUUVVVVWWWHHWXXXXXYYYYYZZZZZ111
GGGGGHHHHHIIIIIJJJJJKKKKKLLLLLMMMMMNNNNNOOOOOPPPPPQQQQQRRRRRSSSSSTTTTTUUUUUVVVVWWWHHWXXXXXYYYYYZZZZZ111
GGGGGHHHHHIIIIIJJJJJKKKKKLLLLLMMMMMNNNNNOOOOOPPPPPQQQQQRRRRRSSSSSTTTTTUUUUUVVVVWWWHHWXXXXXYYYYYZZZZZ111
HHHHHIIIIIJJJJJKKKKKLLLLLMMMMMNNNNNOOOOOPPPPPQQQQQRRRRRSSSSSTTTTTUUUUUVVVVWWWHHWXXXXXYYYYYZZZZZ1111122
HHHHHIIIIIJJJJJKKKKKLLLLLMMMMMNNNNNOOOOOPPPPPQQQQQRRRRRSSSSSTTTTTUUUUUVVVVWWWHHWXXXXXYYYYYZZZZZ1111122
HHHHHIIIIIJJJJJKKKKKLLLLLMMMMMNNNNNOOOOOPPPPPQQQQQRRRRRSSSSSTTTTTUUUUUVVVVWWWHHWXXXXXYYYYYZZZZZ1111122
IIIIIJJJJJKKKKKLLLLLMMMMMNNNNNOOOOOPPPPPQQQQQRRRRRSSSSSTTTTTUUUUUVVVVWWWHHWXXXXXYYYYYZZZZZ11111222223
IIIIIJJJJJKKKKKLLLLLMMMMMNNNNNOOOOOPPPPPQQQQQRRRRRSSSSSTTTTTUUUUUVVVVWWWHHWXXXXXYYYYYZZZZZ11111222223
IIIIIJJJJJKKKKKLLLLLMMMMMNNNNNOOOOOPPPPPQQQQQRRRRRSSSSSTTTTTUUUUUVVVVWWWHHWXXXXXYYYYYZZZZZ11111222223
JJJJJKKKKKLLLLLMMMMMNNNNNOOOOOPPPPPQQQQQRRRRRSSSSSTTTTTUUUUUVVVVWWWHHWXXXXXYYYYYZZZZZ11111222223333344
JJJJJKKKKKLLLLLMMMMMNNNNNOOOOOPPPPPQQQQQRRRRRSSSSSTTTTTUUUUUVVVVWWWHHWXXXXXYYYYYZZZZZ11111222223333344
JJJJJKKKKKLLLLLMMMMMNNNNNOOOOOPPPPPQQQQQRRRRRSSSSSTTTTTUUUUUVVVVWWWHHWXXXXXYYYYYZZZZZ11111222223333344
KKKKKLLLLLMMMMMNNNNNOOOOOPPPPPQQQQQRRRRRSSSSSTTTTTUUUUUVVVVWWWHHWXXXXXYYYYYZZZZZ11111222223333344444555
KKKKKLLLLLMMMMMNNNNNOOOOOPPPPPQQQQQRRPRPSSSSSTTTTTUUUUUVVVVWWWHHWXXXXXYYYYYZZZZZ11111222223333344444555
KKKKKLLLLLMMMMMNNNNNOOOOOPPPPPQQQQQRRRRRSSSSSTTTTTUUUUUVVVVWWWHHWXXXXXYYYYYZZZZZ11111222223333344444555
LLLLLMMMMMNNNNNOOOOOPPPPPQQQQQRRRRRSSSSSTTTTTUUUUUVVVVVWWWHHWXXXXXYYYYYZZZZZ11111222223333344444555556666
LLLLLMMMMMNNNNNOOOOOPPPPPQQQQQRRFFRSSSSSTTTTTUUUULLVVVVVWWWHHWXXXXXYYYYYZZZZZ11111222223333344444555556666
LLLLLMMMMMNNNNNOOOOOPPPPPQQQQQRRRRRSSSSSTTTTTUUUUUVVVVVWWWHHWXXXXXYYYYYZZZZZ11111222223333344444555556666
MMMMMNNNNNOOOOOPPPPPQQQQQRRRRRSSSSSTTTTTUUUUUVVVVVWWWHHWXXXXXYYYYYZZZZZ11111222223333344444555556666677
MMMMMNNNNNOOOOOPPPPPQQQQQRRRRRSSSSSTTTTTUUUUUVVVVVWWWHHWXXXXXYYYYYZZZZZ11111222223333344444555556666677
MMMMMNNNNNOOOOOPPPPPQQQQQRRRRRSSSSSTTTTTUUUUUVVVVVWWWHHWXXXXXYYYYYZZZZZ11111222223333344444555556666677
NNNNNOOOOOPPPPPQQQQQRRRRRSSSSSTTTTTUUUUUVVVVVWWWHHWXXXXXYYYYYZZZZZ11111222223333344444555556666677777888
NNNNNOOOOOPPPPPQQQQQRRRRRSSSSSTTTTTUUUUUVVVVVWWWHHWXXXXXYYYYYZZZZZ11111222223333344444555556666677777888
NNNNNOOOOOPPPPPQQQQQRRRRRSSSSSTTTTTUUUUUVVVVVWWWHHWXXXXXYYYYYZZZZZ11111222223333344444555556666677777888
OOOOOPPPPPQQQQQRRRPPSSSSSTTTTTUUUUUVVVVVWWWHHWXXXXXYYYYYZZZZZ11111222223333344444555556666677777888889999
OOOOOPPPPPQQQQQRRPRRSSSSSTTTTTUUUUUVVVVVWWWHHWXXXXXYYYYYZZZZZ11111222223333344444555556666677777888889999
CCOOOPPPPPQQQQQRPRRRSSSSSTTTTTUUUUUVVVVVWWWHHWXXXXXYYYYYZZZZZ11111222223333344444555556666677777888889999
PPPPPQQQQQRRRRRSSSSSTTTTTUUUUUVVVVVWWWHHWXXXXXYYYYYZZZZZ11111222223333344444555556666677777888889999++++
PPPPPQQQQQRRRRRSSSSSTTTTTUUUUUVVVVVWWWHHWXXXXXYYYYYZZZZZ11111222223333344444555556666677777888889999++++
PPPPPQQQQQRRRRRSSSSSTTTTTUUUUUVVVVVWWWHHWXXXXXYYYYYZZZZZ11111222223333344444555556666677777888889999++++
QQQQQRRRRRSSSSSTTTTTUUUUUVVVVVWWWHHWXXXXXYYYYYZZZZZ11111222223333344444555556666677777888889999++++----
QQQQQRRRRRSSSSSTTTTTUUUUUVVVVVWWWHHWXXXXXYYYYYZZZZZ11111222223333344444555556666677777888889999++++----
QQQQQRRRRRSSSSSTTTTTUUUUUVVVVVWWWHHWXXXXXYYYYYZZZZZ11111222223333344444555556666677777888889999++++----
RRRRRSSSSSTTTTTUUUUUVVVVVWWWHHWXXXXXYYYYYZZZZZ11111222223333344444555556666677777888889999++++----====
RRRRRSSSSSTTTTTUUUUUVVVVVWWWHHWXXXXXYYYYYZZZZZ11111222223333344444555556666677777888889999++++----====
RRRRRSSSSSTTTTTUUUUUVVVVVWWWHHWXXXXXYYYYYZZZZZ11111222223333344444555556666677777888889999++++----====
SSSSSTTTTTUUUUUVVVVVWWWHHWXXXXXYYYYYZZZZZ11111222223333344444555556666677777888889999++++----=====****
SSSSSTTTTTUUUUUVVVVVWWWHHWXXXXXYYYYYZZZZZ11111222223333344444555556666677777888889999++++----=====****
SSSSSTTTTTUUUUUVVVVVWWWHHWXXXXXYYYYYZZZZZ11111222223333344444555556666677777888889999++++----=====****
TTTTTUUUUUVVVVVWWWHHWXXXXXYYYYYZZZZZ11111222223333344444555556666677777888889999++++----=====****:::
TTTTTUUUUUVVVVVWWWHHWXXXXXYYYYYZZZZZ11111222223333344444555556666677777888889999++++----=====****:::
TTTTTUUUUUVVVVVWWWHHWXXXXXYYYYYZZZZZ11111222223333344444555556666677777888889999++++----=====****:::
UUUUUVVVVVWWWHHWXXXXXYYYYYZZZZZ11111222223333344444555556666677777888889999++++----=====****:::::
UUUUUVVVVVWWWHHWXXXXXYYYYYZZZZZ11111222223333344444555556666677777888889999++++----=====****:::::
UUUUUVVVVVWWWHHWXXXXXYYYYYZZZZZ11111222223333344444555556666677777888889999++++----=====****:::::
```

Figure 4.10   Some output from MAP3.

MAP3 and MAP4, groups of five characters must be set in the initialization statement, instead of the single characters used in MAP1 and MAP2. This is done by declaring CHARS in MAP3 and TABLE in MAP4 to be double-length words so that they will accommodate the extra character above the four which will fit in a standard word. Thus, for MAP3

```
REAL*8 CHARS(41) /'AAAAA','BBBBB','CCCCC',.....
```

and for MAP4:

```
REAL*8 TABLE(7,3)/'.....',' +++++',' =====',......
```

The output array for both subroutines (LINE) must be changed to double-length and reduced in size because of the reduced capacity of each page width. If we assume 125 characters in a line, then the declaration for MAP3 is

```
REAL*8 LINE(25)
```

and for MAP4,

```
REAL*8 LINE(25,3)
```

Other changes required in the first part of each subroutine are the subroutine names themselves, and the comment cards which identify and describe the program's function. The computations which follow this first section (area computation and data checking in MAP1, and calculating the maximum, minimum, range, and class interval in MAP2) remain exactly the same in MAP3 and MAP4, so are not presented here.

The most substantial changes to MAP1 and MAP2 are those which allow multiple page widths and which print and overprint blocks of characters in MAP3 and MAP4. Figure 4.11 is a flow chart describing this section in MAP4. The FORTRAN coding for this part of MAP3 and MAP4 is presented below using where possible the same statements that were used in LISTR.

The FORTRAN statements for MAP3 are:

```
C COMPUTE NUMBER OF MAP SECTIONS (PAGE WIDTHS) ASSUMING 25 NUMBERS     MAP3   39
C PER PAGE                                                            MAP3   40
      NSECS=(NCHARS+24)/25                                            MAP3   41
C START THE OUTPUT LOOP                                               MAP3   42
      DO 9 K=1,NSECS                                                  MAP3   43
      KB=K*25                                                         MAP3   44
      KA=KB-24                                                        MAP3   45
C CHECK FOR RIGHT SIDE OF MAP                                         MAP3   46
      IF(KB.GT.NCHARS) KB=NCHARS                                      MAP3   47
C PRINT A TITLE AT THE TOP OF EACH MAP SECTION                        MAP3   48
      WRITE(6,12)K,NSECS                                              MAP3   49
12    FORMAT('1THIS IS MAP SECTION',I3,' OF',I3)                      MAP3   50
      DO 9 J=1,NLINES                                                 MAP3   51
C ASSIGN CHARACTER GROUPS TO LINE                                     MAP3   52
C USE LL TO DESIGNATE LOCATION IN LINE                                MAP3   53
      LL=0                                                            MAP3   54
      DO 13 I=KA,KB                                                   MAP3   55
      LL=LL+1                                                         MAP3   56
13    LINE(LL)=CHARS(X(I,J))                                          MAP3   57
C OUTPUT LINE THREE TIMES                                             MAP3   58
      DO 9 JJ=1,3                                                     MAP3   59
9     WRITE(6,31)(LINE(I),I=1,LL)                                     MAP3   60
31    FORMAT(' ',25A5)                                                MAP3   61
```

The coding for this section in MAP4 is somewhat more complicated because of the overprinting:

```
C COMPUTE NUMBER OF MAP SECTIONS (PAGE WIDTHS) ASSUMING 25 NUMBERS     MAP4   29
C PER PAGE                                                            MAP4   30
      NSECS=(NCHARS+24)/25                                            MAP4   31
C START THE OUTPUT LOOP                                               MAP4   32
      DO 9 K=1,NSECS                                                  MAP4   33
      KB=K*25                                                         MAP4   34
      KA=KB-24                                                        MAP4   35
C CHECK FOR RIGHT SIDE OF MAP                                         MAP4   36
      IF(KB.GT.NCHARS)KB=NCHARS                                       MAP4   37
C PRINT A TITLE AT THE TOP OF EACH MAP SECTION                        MAP4   38
      WRITE(6,12)K,NSECS                                              MAP4   39
12    FORMAT('1THIS IS MAP SECTION',I3,' OF',I3)                      MAP4   40
      DO 9 J=1,NLINES                                                 MAP4   41
C ASSIGN CHARACTER GROUPS TO LINE                                     MAP4   42
C USE LL TO DESIGNATE LOCATION IN LINE                                MAP4   43
      LL=0                                                            MAP4   44
      DO 13 I=KA,KB                                                   MAP4   45
      LL=LL+1                                                         MAP4   46
      KL=(X(I,J)-XMIN)/CLASS+1.                                       MAP4   47
      DO 13 L=1,3                                                     MAP4   48
13    LINE(LL,L)=TABLE(KL,L)                                          MAP4   49
C OUTPUT LINE THREE TIMES                                             MAP4   50
      DO 9 JJ=1,3                                                     MAP4   51
C PRINT                                                               MAP4   52
      WRITE(6,31)(LINE(I,1),I=1,LL)                                   MAP4   53
```

```
31      FORMAT(' ',25A5)                              MAP4   54
C OVERPRINT                                           MAP4   55
        DO 9 L=2,3                                    MAP4   56
9       WRITE(6,53)(LINE(I,L),I=1,LL)                 MAP4   57
53      FORMAT('+',25A5)                              MAP4   58
```

The legend coding from MAP1 and MAP2 can be used in MAP3 and MAP4 with
no change if the user is satisfied with a single character to represent the texture or
grey level pattern. More appropriate, however, would be a block of characters for
each class. The changes required in the original coding to do this are the placement of
the print and overprint WRITE statements within a loop, changing formats from A1

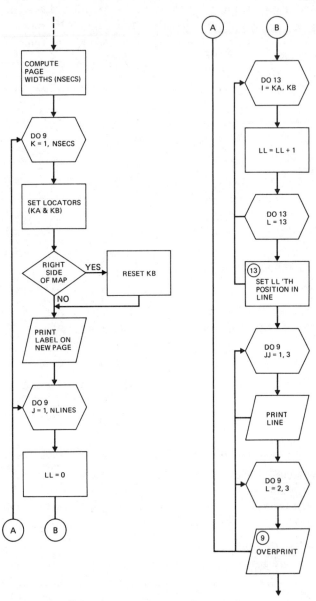

Figure 4.11 Flowchart for central part of MAP4.

to A5, and arranging the output statements so that the class number and limits can be printed with the first line of the character blocks. These changes are contained in the FORTRAN coding below:

```
C                                                                    MAP3  62
C MAP LEGEND                                                         MAP3  63
      WRITE(6,80)                                                    MAP3  64
80    FORMAT('1SYMBOL    TYPE NUMBER    AREA (%)')                   MAP3  65
      DO 90 K=1,41                                                   MAP3  66
C IF A TYPE HAS ZERO AREA, DELETE IT FROM THE LIST                   MAP3  67
      IF(AREA(K).EQ.0.) GO TO 90                                     MAP3  68
C FIRST LINE                                                         MAP3  69
      WRITE(6,100)CHARS(K),K,AREA(K)                                 MAP3  70
100   FORMAT('0',A5,9X,I2,9X,F5.1)                                   MAP3  71
C SECOND AND THIRD LINES                                             MAP3  72
      DO 101 L=1,2                                                   MAP3  73
101   WRITE(6,102)CHARS(K)                                          MAP3  74
102   FORMAT(' ',A5)                                                MAP3  75
90    CONTINUE                                                       MAP3  76
      RETURN                                                         MAP3  77
      END                                                            MAP3  78
```

THIS IS MAP SECTION 1 OF 1

$$\begin{aligned}
&\text{SSSSS} //// \text{XXXXXXXXXX} \text{SSSSS} \text{HHHHHHHHHHH} //// \text{XXXXXXXXXX}\\
\end{aligned}$$

Figure 4.12   Some output from MAP4. (Part one.)

```
C LEGEND STARTS ON NEW PAGE WITH TITLE                                    MAP4  59
      WRITE(6,100)                                                         MAP4  60
100   FORMAT('1CLASS NUMBER   GRAY LEVEL   LOWER LIMIT   UPPER LIMIT')     MAP4  61
C RESET CLASS INTERVAL TO EXACT VALUE                                     MAP4  62
      CLASS=RANGE/7.                                                      MAP4  63
C GO THROUGH LOOP SEVEN TIMES, ONCE FOR EACH CLASS                        MAP4  64
      DO 120 K=1,7                                                        MAP4  65
C SET CLASS LIMITS                                                        MAP4  66
      BOT=XMIN+CLASS*(K-1)                                                MAP4  67
      TOP=BOT+CLASS                                                       MAP4  68
C FIRST LINE                                                              MAP4  69
C PRINT                                                                   MAP4  70
      WRITE(6,110) K,TABLE(K,1),BOT,TOP                                  MAP4  71
110   FORMAT('0',5X,I2,5X,A5,7X,F11.2,2X,F11.2)                          MAP4  72
C OVERPRINT                                                               MAP4  73
      DO 105 L=2,3                                                        MAP4  74
105   WRITE(6,115) TABLE(K,L)                                            MAP4  75
115   FORMAT('+',12X,A5)                                                 MAP4  76
C SECOND AND THIRD LINES                                                  MAP4  77
      DO 120 JJ=1,2                                                       MAP4  78
C PRINT                                                                   MAP4  79
      WRITE(6,125) TABLE(K,1)                                            MAP4  80
125   FORMAT(' ',12X,A5)                                                 MAP4  81
C OVERPRINT                                                               MAP4  82
      DO 120 L=2,3                                                        MAP4  83
120   WRITE(6,115) TABLE(K,L)                                            MAP4  84
      RETURN                                                             MAP4  85
      END                                                                MAP4  86
```

CLASS NUMBER	GRAY LEVEL	LOWER LIMIT	UPPER LIMIT
1	....	-1.95	-1.39
2	+++++ +++++ +++++	-1.39	-0.84
3	===== ===== =====	-0.84	-0.28
4	XXXXX XXXXX XXXXX	-0.28	0.28
5	///// ///// /////	0.28	0.84
6	$$$$$ $$$$$ $$$$$	0.84	1.39
7	HHHHH HHHHH HHHHH	1.39	1.95

Figure 4.12   (Part two.)

4.8   Example 9. SQUEZ1 and SQUEZ2: Preparing Matrix Data for MAP1 and MAP2

When MAP3 and MAP4 are used with larger data sets, the machine produces several pages of output, which then must be trimmed and taped together. The result is quite a large map, ideal for presentation to a group, but unwieldy for the individual when compared with the single character maps of MAP1 and MAP2. This example describes a procedure by which matrices digitized with a square grid can be modified to a form based on a rectangular grid so that they may be mapped using MAP1 or MAP2.

SQUEZ1 and SQUEZ2 reduce the number of columns in the data matrix to three-fifths of their original number by forming new columns, each consisting of five-thirds old columns.[7] Each element in this new matrix will thus correspond to a rectangular area on the source map, with a ratio of sides such that it can be represented by a single character location on a computer output page. The cost to the technique is obvious: information in the original data matrix is lost in the aggregation procedure.

```
$#XX$BB#XX#BB$XX#BB$#X#$B          B#XX$BB#XXSBB$XX#BB$#X#$B
B#XX$BB#XX$BB$XX#BB$#X#$B          #=+=#$$X==X$$X=+X#$#=+=#$
#X==#$$X==X$$#==X#$#===#$          =+..=XX+..+XX=..+XX=...=X
X+.+=XX+..=XX=..+XX=+.+=X          X+.+=#X+...=XX=..+X#=+.+X#
=+..=XX+..+XX=..+=X=...=X          $X=X$B$#===#BB#X=X$B$X=X$B
X=++X##=++X##X++=##X=+=X#          $X=X$BB#XX#BB#X=#$BB$X=X$B
$X=X$B$#XX#BB#X=#$BB$X=X$B         X+..=XX+..=XX=..+XX=+.+=X
B#X#$BB#XX$BB$XX#BB$#X#$B          X+.+X##=.+=##=+.=X#X+.+X#
$X=X#$$X==#$$#==X$$#X=X#$          $#XX$BB#XX#BB$XX#BB$XXX$B
X+.+X#X=.+=##=+.=X#X+.+X#          $X==#$$X==#$$#==X$$#==X#$
=+..=XX+..+XX+..+=X=...=X          X+..=XX+..=XX=..+XX=+.+=X
X+.+X#X=.+=##=+.=X#X+.+X#          #=++X##=++X##X++=##X=+=X#
$X=X#$$X==#$$#==X$$#X=X#$          B#XX$BB#XX$BB$XX#BB$#X#$B
B#X#$BB#XX$BB$XX#BB$#X#$B          #=+=#$#X++X$$X=+=#$#=+=#$
$X=X$B$#XX#BB#X=#$BB$X=X$B         =+..=XX+..+XX=..+XX=...=X
X=++X##=++X##X++=##X=+=X#
=+..=XX+..+XX=..+=X=...=X
X+.+=XX+..=XX=..+XX=+.+=X
#X==#$$X==X$$#==X#$#===#$
B#XX$BB#XX$BB$XX#BB$#X#$B
$#XX$BB#XX#BB$XX#BB$#X#$B
#=+=X$#X+=X#$X=+X#$#=+=#$
=+..=XX+..=XX=..+XX=+.+=X
=+..=XX+..+XX=..+XX=...=X
#=+=X##X++X##X++=#$X=+=#$
```

Figure 4.13 An illustration of the effect of SQUEZ2. The output on the left was produced by MAP2 from a 25 by 25 matrix. The output on the right was produced by processing the same 25 by 25 matrix with SQUEZ2, then mapping it with MAP2. (The matrix was computed by setting each element to the sum of the cosine of its row number and the sine of its column number.)

The computation procedure for this collapsing is similar for qualitative maps (SQUEZ1) and quantitative maps (SQUEZ2). The value of each element in the first column of the new (collapsed) matrix is derived from the corresponding element in the first column of the original matrix together with two-thirds of each element in the

---

[7] The usual assumption is made about a six line per inch printer.

second column of the original matrix. The second new column is derived from one-third of column 2, all of column 3, and one-third of column 4 in the original matrix. The third new column is derived from two-thirds of column 4 and all of column 5. This pattern then repeats itself for each five columns of the original matrix and three columns in the new matrix.

The assignment statements which set values in the new matrix depend upon the type of map. In SQUEZ1 (qualitative data), only the first, third, and fifth of each five columns in the original data matrix need be considered in determining each three columns in the new matrix. This is because the second and fourth columns contribute only fractional amounts (one-third or two-thirds) in the aggregation totals. Any information stored in these columns is completely lost in the collapsed map. The FORTRAN statements required are of the form:

```
XNEW(I,J)=XOLD(I,JJ)
XNEW(I,J+1)=XOLD(I,JJ+2)
XNEW(I,J+2)=XOLD(I,JJ+4)
```

where XNEW is the collapsed matrix for mapping, XOLD is the original data matrix, *J* is defined to have the values 1, 4, 7, 10, etc. (every third column in XNEW), and *JJ* is defined to have the values 1, 6, 11, 16, etc. (every fifth column in XOLD).

The assignment statements required for quantitative data (SQUEZ2) take into account all five columns in the original map for each three in the new map by an averaging process which uses the contribution of each column in the original matrix as a weight:

```
XNEW(I,J)=(XOLD(I,JJ)+2./3.*XOLD(I,JJ+1))*3./5.
XNEW(I,J+1)=(1./3.*XOLD(I,JJ+1)+XOLD(I,JJ+2)+1./3.*XOLD(I,JJ+3))
X     *3./5.
XNEW(I,J+2)=(2./3.*XOLD(I,JJ+3)+XOLD(I,JJ+4))*3./5.
```

If the data are counts or area measures rather than elevations on a surface, then totals rather than averages are required, and the weight (3./5.) can be eliminated. (Including the weight, however, affects only the legend, not the map.)

In both SQUEZ1 and SQUEZ2, the three assignment statements are placed within a DO loop which is executed as many times as there are cycles of three columns in the new matrix. If there are any columns left over, they are considered in special assignment statements outside the loop.

Let us now consider the FORTRAN coding for SQUEZ1. It starts with the usual descriptive comment cards and declarations. Array space is required for an input matrix (called XOLD above) and the collapsed matrix (XNEW above). However, considerable memory space can be saved if one matrix is used for both data and collapsed matrix. This is possible in this particular problem because each column in the new matrix can be stored in the original matrix without destroying any values required for calculating subsequent columns. The original data matrix is destroyed, however. This means that if the data are needed for further calculations in the calling program after the mapping step, then the original data must be saved in another array space before SQUEZ1 is called. (This procedure is illustrated in a sample main program used with SQUEZ2 later in this section.)

The initial statements in SQUEZ1 are as follows:

```
      SUBROUTINE SQUEZ1(MAP,NROWS,NCOLS,KCOLS)            SQZ1    1
C PREPARES DATA FOR INPUT TO MAP1                         SQZ1    2
C MAP CONTAINS ORIGINAL DATA MATRIX, NROWS BY NCOLS       SQZ1    3
C WHEN SUBROUTINE CALLED, AND                             SQZ1    4
```

```
C COLLAPSED MATRIX, NROWS BY KCOLS, ON RETURN          SQZ1   5
        INTEGER MAP(NROWS,NCOLS),REMAIN                 SQZ1   6
```

The first assignment statement calculates now many columns will be in the reduced map:

```
    KCOLS=(NCOLS*3)/5                                   SQZ1   7
```

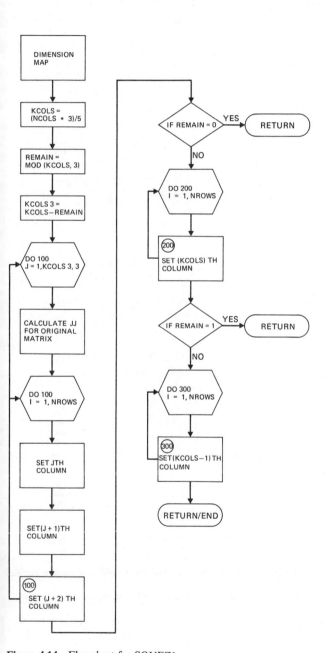

Figure 4.14   Flowchart for SQUEZ1.

The outermost DO statement in the main loop uses an increment value of 3 in order to obtain a sequence of DO variable values which identify the first column in the new matrix for each cycle, that is, 1, 4, 7, 10, 13, 16, ... The test value in this DO statement must be set to that value which is less than or equal to KCOLS and which is exactly divisible by three. Otherwise, subscripts in later assignment statements ($JJ+2$ and $JJ+4$ on cards 14 and 15 will refer to undefined columns of MAP. This is done with a FORTRAN-supplied function for modulo or remainder arithmetic. The statement

```
      REMAIN=MOD(KCOLS,3)                                     SQZ1   8
```

stores in REMAIN the remainder after division of KCOLS by 3. (AMOD is the remainder function for real arguments.) The test value of the loop is set by

```
      KCOLS3=KCOLS-REMAIN                                     SQZ1   9
```

and the loop started:

```
      DO 100 J=1,KCOLS3,3                                     SQZ1  10
```

For each value of *J* designating a column in the new matrix, it is necessary to calculate the number of the corresponding column in the original matrix:

```
      JJ=((J-1)*5.)/3.+1.                                     SQZ1  11
```

JJ will thus have the sequence of values 1, 6, 11, 16, 21, . . . , as *J* goes 1, 4, 7, 10, 13, ...

The inner DO statement refers to rows, and is followed by the three assignment statements:

```
      DO 100 I=1,NROWS                                        SQZ1  12
      MAP(I,J)=MAP(I,JJ)                                      SQZ1  13
      MAP(I,J+1)=MAP(I,JJ+2)                                  SQZ1  14
100   MAP(I,J+2)=MAP(I,JJ+4)                                  SQZ1  15
```

This loop will result in a new matrix of exactly KCOLS only when REMAIN equals 0. If it equals 1 or 2 (the only other possible values), additional coding is required to compute the extra columns to bring the matrix to KCOLS:

```
C IF NO REMAINDER, MATRIX COMPLETE                           SQZ1  16
      IF(REMAIN.EQ.0) RETURN                                 SQZ1  17
C SET LAST COLUMN                                            SQZ1  18
      DO 200 I=1,NROWS                                        SQZ1  19
200   MAP(I,KCOLS)=MAP(I,NCOLS)                               SQZ1  20
C CHECK IF A SECOND LAST COLUMN REQUIRED                     SQZ1  21
      IF(REMAIN.EQ.1) RETURN                                 SQZ1  22
C SET SECOND LAST COLUMN                                     SQZ1  23
      DO 300 I=1,NROWS                                        SQZ1  24
300   MAP(I,KCOLS-1)=MAP(I,NCOLS-2)                           SQZ1  25
```

Control is then returned to the calling program:

```
      RETURN                                                 SQZ1  26
      END                                                    SQZ1  27
```

The map is produced by calling MAP1 from the main program, using the values of KCOLS as an argument, for example:

```
      CALL MAP1(MAP,NROWS,KCOLS)
```

SQUEZ2 differs from SQUEZ1 in three ways:
1 the matrix is real and named X;
2 three variables (ONETHR, TWOTHR and THRFIF are set before the loop starts;

3   the assignment statements which set values in the new matrix are weighted
    averages.

The second of these differences illustrates once again an important programming
practice: no computations are included within a DO loop which can be completed
before its initiation. In the three assignment statements presented in the initial discussion of SQUEZ2, the fractions 1./3., 2./3., and 3./5. were included in the
arithmetic expressions. If these statements were included in this form in the final
version of SQUEZ2, the total number of calculations would be almost twice as many
as in the version where the fractions are set beforehand. (However, if the fraction
5./3. were set before the loop, the resulting calculations would be in error; the reader
is invited to determine why.)

The FORTRAN statements for SQUEZ2 are presented below together with a
sample main program which illustrates how a data array may be saved before
SQUEZ2 is called.

```
C MAIN PROGRAM TO READ TWO MATRICES, MAP EACH, THEN SUM THEM      MAIN   1
C ELEMENT BY ELEMENT, AND MAP THIS                                MAIN   2
      DIMENSION X1(10,16),X2(10,16),DATA(10,16)                   MAIN   3
C READ FIRST MAP INTO X1                                          MAIN   4
      DO 10 J=1,16                                                MAIN   5
10    READ(5,100)(X1(I,J),I=1,10)                                 MAIN   6
C PUT X1 INTO DATA TO SEND TO SQUEZ2                              MAIN   7
      DO 20 J=1,16                                                MAIN   8
      DO 20 I=1,10                                                MAIN   9
20    DATA(I,J)=X1(I,J)                                           MAIN  10
      CALL SQUEZ2(DATA,10,16,KCOLS)                               MAIN  11
      CALL MAP2(DATA,10,KCCLS)                                    MAIN  12
C READ SECOND MAP                                                 MAIN  13
      DO 30 J=1,16                                                MAIN  14
30    READ(5,100)(X2(I,J),I=1,10)                                 MAIN  15
C PUT X2 INTO DATA TO SEND TO SQUEZ2                              MAIN  16
      DO 40 J=1,16                                                MAIN  17
      DO 40 I=1,10                                                MAIN  18
40    DATA(I,J)=X2(I,J)                                           MAIN  19
      CALL SQUEZ2(DATA,10,16,KCCLS)                               MAIN  20
      CALL MAP2(DATA,10,KCOLS)                                    MAIN  21
C SUM X1 AND X2 IN DATA                                           MAIN  22
      DO 50 J=1,16                                                MAIN  23
      DO 50 I=1,10                                                MAIN  24
50    DATA(I,J)=X1(I,J)+X2(I,J)                                   MAIN  25
      CALL SQUEZ2(DATA,10,16,KCCLS)                               MAIN  26
      CALL MAP2(DATA,10,KCCLS)                                    MAIN  27
      STOP                                                        MAIN  28
100   FORMAT(10F5.0)                                              MAIN  29
      END                                                         MAIN  30


      SUBROUTINE SQUEZ2(X,NROWS,NCOLS,KCOLS)                      SQZ2   1
C PREPARES DATA FOR INPUT TO MAP2                                 SQZ2   2
C X CONTAINS ORIGINAL DATA MATRIX, NROWS BY NCOLS                 SQZ2   3
C WHEN SUBROUTINE CALLED, AND                                     SQZ2   4
C COLLAPSED MATRIX, NROWS BY KCOLS, ON RETURN                     SQZ2   5
      DIMENSION X(NROWS,NCOLS)                                    SQZ2   6
      INTEGER REMAIN                                              SQZ2   7
C SET CONSTANTS                                                   SQZ2   8
      ONETHR=1./3.                                                SQZ2   9
      TWOTHR=2./3.                                                SQZ2  10
      THRFIF=3./5.                                                SQZ2  11
      KCOLS=(NCOLS*3)/5                                           SQZ2  12
      REMAIN=MOD(KCOLS,3)                                         SQZ2  13
      KCOLS3=KCOLS-REMAIN                                         SQZ2  14
      DO 100 J=1,KCOLS3,3                                         SQZ2  15
      DO 100 I=1,NROWS                                            SQZ2  16
      JJ=((J-1)*5.)/3.+1.                                         SQZ2  17
      WRITE(6,999) J,JJ                                           SQZ2  18
999   FORMAT(' ',2I5)                                             SQZ2  19
      X(I,J)=(X(I,JJ)+TWOTHR*X(I,JJ+1))*THRFIF                    SQZ2  20
```

```
      X(I,J+1)=(ONETHR*X(I,JJ+1)+X(I,JJ+2)+ONETHR*X(I,JJ+3))*THRFIF      SQZ2  21
100   X(I,J+2)=(TWOTHR*X(I,JJ+3)+X(I,JJ+4))*THRFIF                        SQZ2  22
C IF NO REMAINDER, MATRIX COMPLETE                                       SQZ2  23
      IF(REMAIN.EQ.0) RETURN                                             SQZ2  24
C FILL LAST COLUMN USING ALL OF LAST COLUMN IN ORIGINAL, AND TWO-THIRDS  SQZ2  25
C OF SECOND LAST COLUMN IN ORIGINAL                                      SQZ2  26
      DO 110 I=1,NROWS                                                   SQZ2  27
110   X(I,KCOLS)=(X(I,NCOLS)+TWOTHR*X(I,NCOLS-1))*THRFIF                  SQZ2  28
C CHECK IF A SECOND LAST COLUMN REQUIRED                                 SQZ2  29
      IF(REMAIN.EQ.1)RETURN                                              SQZ2  30
C FILL SECOND LAST COLUMN USING ONE THIRD OF SECOND LAST COLUMN IN       SQZ2  31
C ORIGINAL ALL OF THIRD LAST, AND ONE THIRD OF FOURTH LAST               SQZ2  32
      DO 120 I=1,NROWS                                                   SQZ2  33
120   X(I,KCOLS-1)=(ONETHR*X(I,NCOLS-1)+X(I,NCOLS-2)+ONETHR*X(I,NCOLS    SQZ2  34
     $  -3))*THRFIF                                                      SQZ2  35
      RETURN                                                             SQZ2  36
      END                                                                SQZ2  37
```

4.9   Class Intervals on Maps of Quantitative Data

In MAP2 and MAP4, the line printer mapping subroutines for quantitative data presented earlier in this chapter, the correspondence between the value of each matrix element and the required overprint characters was determined with an assignment statement such as

```
NUM=(X(I,J)-XMIN)/CLASS+1.
```

where NUM is the number of the grey level pattern (and hence the printer characters), X the data, XMIN the minimum value of the data, and CLASS the size of one class (in fact, just slightly larger than one class interval to avoid problems with the maximum). This procedure does not always produce satisfactory maps for those data sets with values concentrated in certain grey level classes. Figure 4.15, a slope map, is typical: the frequency distribution of its data values is so strongly skewed to lower values that most of the map is light.

Varying class intervals can significantly change the appearance of maps compiled from skewed distributions or those with strong central tendency. The upper part of figure 4.16 illustrates the result of dividing the range of a distribution into equal class intervals. One third of the map area will be in the central two classes, and approximately six per cent in the extreme classes. The lower part of this figure illustrates a set of class limits which provide more detail in those areas from the centre of the distribution, with a cost in less detail in areas from the extremes. Each of the ten classes has one tenth the data values, and therefore one-tenth the map area.

There are three ways in which class intervals can be varied for a particular computer mapping problem, as follows:

1   require the user to specify the class limits as part of his data;
2   automatically scale the data into classes which have equal numbers of values;
3   transform the data to reduce skewness or central tendency, then scale by the usual method.

The first of these alternatives is suggested as a problem at the end of the chapter. The second is considered in the next section, and the third is described in the section following that.

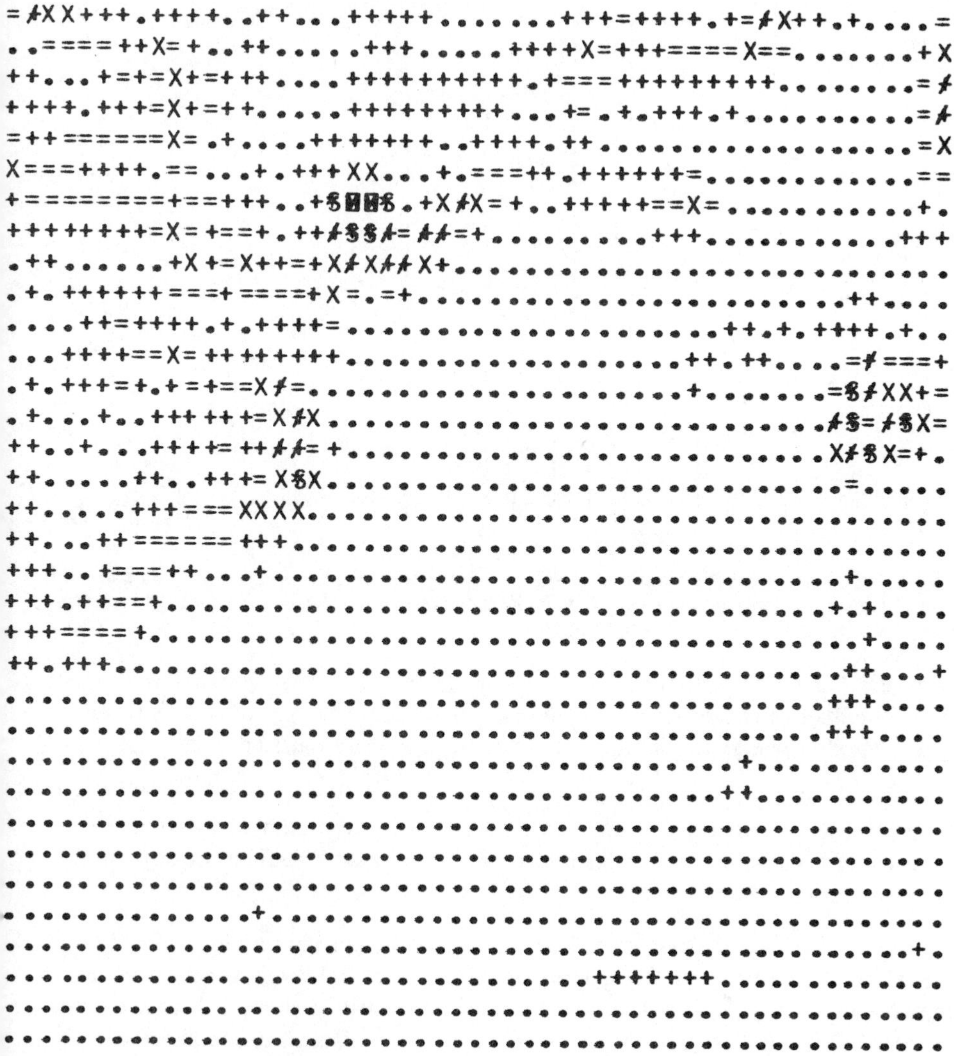

Figure 4.15   A slope map produced by SQUEZ 2 and MAP2.

4.10   Example 10. MAP2A and SORT: Equal Areas of Grey Level Patterns on Quantitative Maps

The method by which data are scaled into classes which have equal numbers is considerably more complicated than dividing the range into equal intervals. It consists of three main steps:

1 sort the data into a vector in which values are arranged in ascending order;
2 determine for the number of grey levels desired those class limits which divide the vector into sections of equal length by

$$L_k = A_{((k-1)^* n/m) + 1}$$
$$U_k = A_{(k^* n/m)}$$

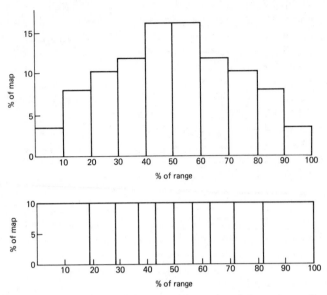

Figure 4.16   Histograms illustrating the results of alternative scaling procedures on a data set with central tendency. Upper: scaling for constant class intervals; lower: scaling for constant class frequencies (map areas).

where L is the lower bound and U the upper bound of class (grey level) k, m is the number of classes, and n the number of values in the vector A. If n is not exactly divisible by m, the upper limit of the mth class must be reset:

$$U_m = A_n$$

3   replace the original data values with a grey level (class) number by comparing each with the upper and lower limits determined in step 2, then proceed with mapping as in MAP2.

The most critical of these steps is the first, because sorting requires so many operations that a program for spatial data could be prohibitively expensive. Once the matrix values are transferred to a vector by concatenating columns, there are several ways to achieve this sorting. A simple and apparently efficient technique consists of determining the minimum value in an increasingly smaller part of the vector, and moving this into the first (sorted) part. FORTRAN coding which will do this is as follows:

```
        XMIN=A(1)
        DO 10 JJ=1,N
        DO 20 K=JJ,N
        IF(A(K).GT.XMIN) GO TO 20
        XMIN=A(K)
        M=K
20      CONTINUE
        II=JJ+1
        L=M+1
        DO 30 J=II,M
        L=L-1
30      A(L)=A(L-1)
10      A(JJ)=XMIN
```

The number of computer operations required by this coding (IF and assignment statements) will be well above N^2. This means that maps which require more than a standard computer page (60 lines by 130 characters) will require hundreds of millions of operations for the sorting! This compares with hundreds of thousands of operations for the same map sizes using MAP2. Clearly a faster procedure is required if MAP2A is to be practical.

Sorting is such an important computer activity that considerable effort has been given to the search for more efficient algorithms, and several have been published which are significantly faster than the procedure described above. The algorithm selected for this example (Shell, 1959; Boothroyd, 1963) operates by comparing pairs of values, and exchanging them if required. It is not the fastest available, but requires the fewest statements, and is easily written in a more general form to sort the columns of a matrix so that the values in a specified row are in ascending order. For example, a three by five matrix

16	26	98	54	8
56	32	78	6	14
9	19	88	34	20

may be sorted using the values in the second row, resulting in the matrix

54	8	26	16	98
6	14	32	56	78
34	20	19	9	88

The more general sorting subroutine is required in two examples presented in the following chapter. It can also be used with MAP2A, which requires only that a vector be sorted, because the subroutine treats the vector as a matrix of one row. (It can be specified either as a vector or a matrix of one row in the calling program; both will have the same effect.) The FORTRAN coding for SORT is presented below, and a flow chart in figure 4.17:

```
      SUBROUTINE SORT(X,N,U,V)                                    SORT    1
C SORTING ROUTINE BASED ON ALGORITHM 201, ASSOCIATION FOR        SORT    2
C COMPUTING MACHINERY, BY J. BOOTHROYD                           SORT    3
C SORTS ROWS OF X(U,N) SO THAT ELEMENTS OF ROW V ARE IN ASCENDING SORT   4
C ORDER.   MAXIMUM OF 25 ROWS                                    SORT    5
      INTEGER U,V,W                                              SORT    6
      REAL X(U,N),SAVE(25)                                       SORT    7
      M=N                                                        SORT    8
100   M=M/2                                                      SORT    9
      IF(M.EQ.0)RETURN                                           SORT   10
      K=N-M                                                      SORT   11
      DO 400 J=1,K                                               SORT   12
      I=J                                                        SORT   13
200   L=I+M                                                      SORT   14
      IF(X(V,I).LE.X(V,L)) GO TO 400                             SORT   15
      DO 300 W=1,U                                               SORT   16
      SAVE(W)=X(W,I)                                             SORT   17
      X(W,I)=X(W,L)                                              SORT   18
300   X(W,L)=SAVE(W)                                             SORT   19
      I=I-M                                                      SORT   20
      IF(I.GT.0) GO TO 200                                       SORT   21
400   CONTINUE                                                   SORT   22
      GO TO 100                                                  SORT   23
      END                                                        SORT   24
```

The sequence of operations in SORT consists of a series of passes through a matrix X, the number of passes depending upon statement 100 (card SORT 9), in which an integer variable M is re-defined repeatedly by halving its previous value and trun-

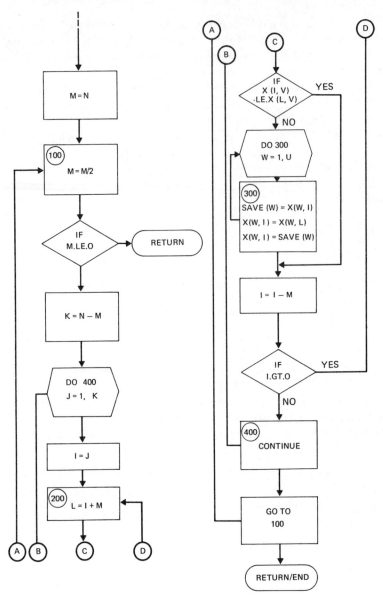

Figure 4.17   Flowchart for SORT.

cating (because integer arithmetic takes place). On each pass, pairs of elements in the specified row (row V) are compared and, if one is larger than the other, they and other elements in the two columns are exchanged in position.

The key to the algorithm is the determination of which pairs of points are to be compared. On each pass, all pairs of elements *M* places apart are compared, plus points which are multiples of *M* places apart, as determined by the statements on cards SORT 20 and SORT 21.

Table 4.2 illustrates the operation of SORT for a matrix one row by nine columns (i.e., a vector of nine elements). The upper part of the table presents the column subscripts of the elements which are compared at each pass, with those which are

Table 4.2 The Operation of SORT with a
Matrix One Row by Nine Columns

	Pass one		Pass two		Pass three	
$M=$	4		2		1	
$K=$	5		7		8	
Subscripts of pairs of values compared:						
	1	5	1	3	1	2
	2	6	2	4	2	3
	3	7	3	5	1	3*
	4	8	1	5*	3	4
	5	9	4	6	2	4*
	1	9*	2	6*	1	4*
			5	7*	4	5*
			3	7*	3	5*
			1	7*	2	5*
			6	8	1	5*
			4	8*	5	6
			2	8*	4	6*
			7	9	3	6*
			5	9*	2	6*
			3	9*	1	6*
			1	9*	6	7
					5	7*
					4	7*
					3	7*
					⋮	⋮

Organization of Data:
Before SORT:
   16 4 7 21 3   2 12 20   2
After pass one:
    3 2 7 20 2   4 12 21 16
After pass two:
    3 2 2   4 7 20 12 21 16
After pass three:
    2 2 3   4 7 12 16 20 21

* Pairs of values which are multiples of M places apart.

multiples of M places apart marked by asterisks. The lower part of the table presents the organization of the data before sorting, and after each pass.

SORT is called by MAP2A, which also determines grey level numbers for each element of the data matrix, prints and overprints appropriate character combinations, and produces a legend. The FORTRAN coding for MAP2A starts with a block of comment cards and declarations:

```
      SUBROUTINE MAP2A(X,NCHARS,NLINES)                              MP2A   1
C PRINTS MAP OF SEVEN GRAY LEVELS OF EQUAL AREA                      MP2A   2
C INPUT DATA (X) IS MATRIX OF REAL NUMBERS NCHARS BY NLINES WHICH IS MP2A   3
C THE TRANSPOSE OF THE LINE PRINTER MAP TO BE PRODUCED               MP2A   4
```

```
C WIDTH OF MAP(NCHARS).LE.131 (ONE PAGE WIDTH ON STANDARD PRINTER)    MP2A   5
C NO RESTRICTION ON LENGTH OF MAP (NLINES) EXCEPT THAT                MP2A   6
C NCHARS*NLINES.LE.5000, THE SIZE OF S, A VECTOR USED FOR SORTING     MP2A   7
C LINE IS USED TO STORE PRINT AND OVERPRINT CHARACTERS FOR ONE LINE OF MP2A  8
C OUTPUT                                                              MP2A   9
C LIMITS CONTAINS CLASS BOUNDS                                        MP2A  10
C TABLE CONTAINS CHARACTERS                                           MP2A  11
      DIMENSION X(NCHARS,NLINES),S(5000)                              MP2A  12
      REAL LINE(131,3)                                                MP2A  13
      REAL LIMITS(7,2)                                                MP2A  14
      REAL TABLE(7,3)/'.','+','=','X','V','S','M',                    MP2A  15
     *               ' ',' ',' ',' ','A','*','E',                     MP2A  16
     *               ' ',' ',' ',' ',' ',' ','W'/                     MP2A  17
```

S is the name of the vector in which the values from the data array X are sorted. Its maximum length has been set at 5000 in the first DIMENSION statement. This also sets upper limits on the size of the data matrix, in that the product of rows and columns (NCHARS*NLINES) must be equal to or less than 5000. This limit may be altered either by changing the DIMENSION statement, or including S and its size as arguments in the SUBROUTINE statement, and giving it an adjustable dimension in the subroutine. The latter procedure is clearly more desirable if the subroutine is to be kept as general as possible. However, it places an additional requirement on the user of MAP2A. He must include in a specification statement and argument list an array which is used only within MAP2A, and for a purpose which may be of no interest to him. For this reason, MAP2A and following examples with the same problem have arrays which are used only within a subroutine dimensioned within that coding.

The data values are assigned to S using a counter K, and then sent to SORT:

```
C PUT X INTO S                                                       MP2A  18
      K=0                                                            MP2A  19
      DO 10 J=1,NLINES                                               MP2A  20
      DO 10 I=1,NCHARS                                               MP2A  21
      K=K+1                                                          MP2A  22
10    S(K)=X(I,J)                                                    MP2A  23
C SORT S                                                             MP2A  24
      CALL SORT(S,K,1,1)                                             MP2A  25
```

The class limits are determined from S using the expressions presented at the beginning of this section:

```
C DETERMINE LIMITS                                                   MP2A  26
      CL=K/7.                                                        MP2A  27
      DO 20 L=1,7                                                    MP2A  28
      LIMITS(L,1)=S((L-1)*CL+1)                                      MP2A  29
20    LIMITS(L,2)=S(L*CL)                                            MP2A  30
```

The upper bound of the last class is reset to allow for errors introduced when the number of data values is not exactly divisible by 7, the number of classes:

```
C RESET UPPER BOUND OF CLASS 7 TO INCLUDE EXTRA VALUES               MP2A  31
      LIMITS(7,2)=S(K)                                               MP2A  32
```

The remainder of the coding of MAP2A is very similar to that of MAP2 except for the procedure by which grey level numbers are determined. In MAP2, this is an assignment statement. In MAP2A, it is a loop containing an IF statement, followed by an assignment statement. The coding for the remainder of MAP2A follows:

```
C POSITION PRINTER AT TOP OF NEW PAGE                                MP2A  33
      WRITE(6,79)                                                    MP2A  34
79    FORMAT('1')                                                    MP2A  35
      DO 9 J=1,NLINES                                                MP2A  36
      DO 13 I=1,NCHARS                                               MP2A  37
C ASSIGN GRAY LEVEL NUMBERS                                          MP2A  38
      DO 40 L=1,7                                                    MP2A  39
      IF(X(I,J).GE.LIMITS(L,1).AND.X(I,J).LE.LIMITS(L,2))GO TO 30    MP2A  40
```

```
40       CONTINUE                                              MP2A   41
C ASSIGN CHARACTERS TO LINE                                    MP2A   42
30       DO 13 K=1,3                                           MP2A   43
13       LINE(I,K)=TABLE(X(I,J),K)                             MP2A   44
C PRINT                                                        MP2A   45
         WRITE(6,31) (LINE(I,1),I=1,NCHARS)                    MP2A   46
31       FORMAT(' ',131A1)                                     MP2A   47
C OVERPRINT                                                    MP2A   48
         DO 9 K=2,3                                            MP2A   49
9        WRITE(6,53)(LINE(I,K),I=1,NCHARS)                     MP2A   50
53       FORMAT('+',131A1)                                     MP2A   51
C LEGEND STARTS ON NEW PAGE WITH TITLE                         MP2A   52
         WRITE(6,100)                                          MP2A   53
100      FORMAT('1CLASS NUMBER  GRAY LEVEL  LOWER BOUND  UPPER BOUND')  MP2A   54
C GO THROUGH LOOP SEVEN TIMES, ONCE FOR EACH CLASS             MP2A   55
         DO 105 K=1,7                                          MP2A   56
C PRINT                                                        MP2A   57
         WRITE(6,110) K,TABLE(K,1),LIMITS(K,1),LIMITS(K,2)     MP2A   58
110      FORMAT('0',5X,I2,9X,A1,7X,F11.2,2X,F11.2)             MP2A   59
```

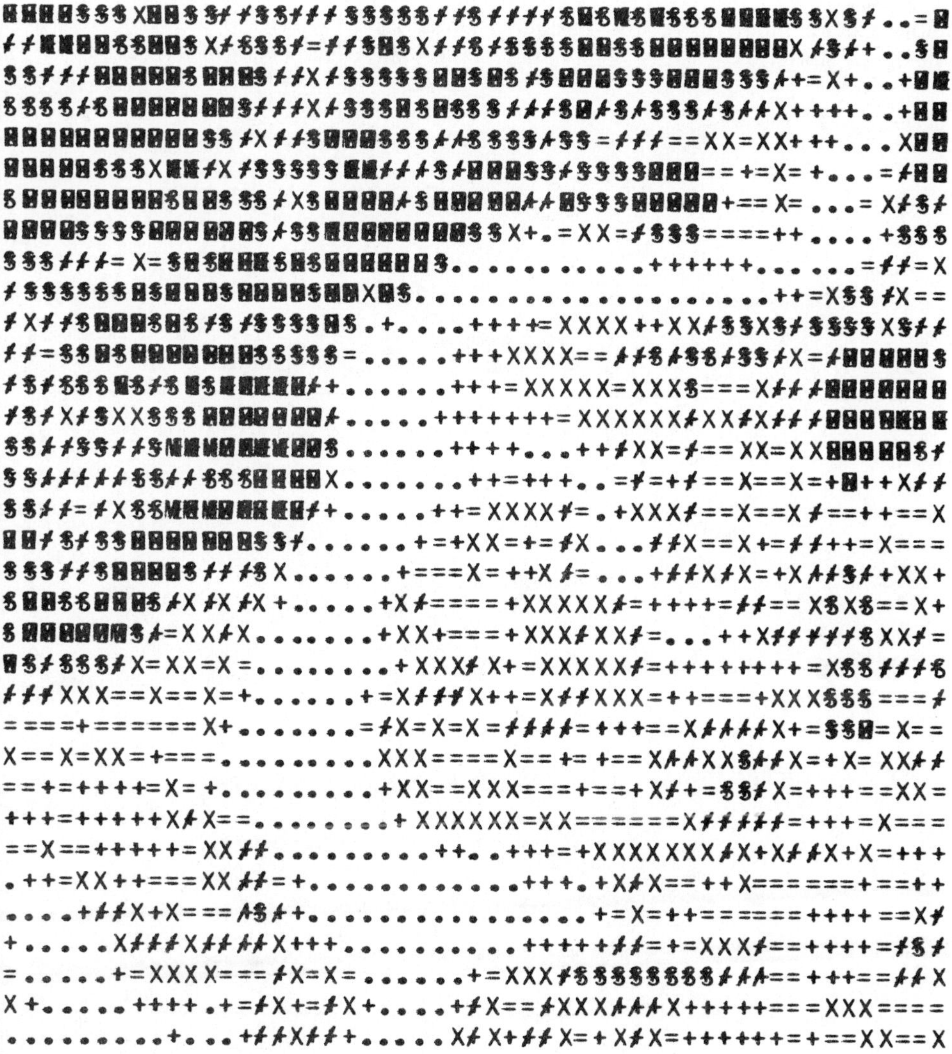

Figure 4.18. A slope map produced by MAP2A.

```
C  OVERPRINT                                    MP2A  60
       DO 105 L=2,3                             MP2A  61
105    WRITE(6,115) TABLE(K,L)                  MP2A  62
115    FORMAT('+',16X,A1)                       MP2A  63
       RETURN                                   MP2A  64
       END                                      MP2A  65
```

4.11   Data Transformations

As efficient as it is, MAP2A is still much slower than dividing the range of the data into equal class intervals (as in MAP2 and MAP4). As a result, it is often useful to attempt an approximation of equal areas of grey on the map by mathematically transforming the data, then applying the conventional scaling procedure. This approach is usually successful for data distributions which are skewed to the upper or lower ends of their ranges. The simplest technique is to replace each data value with either the square root or the logarithm of the absolute value of the difference between it and either the minimum (if the distribution is skewed to low values), or the maximum (if the distribution is skewed to high values), plus one. In mathematical terms:

$$x = \sqrt{x_s - x' + 1}$$

or

$$x = \log_{10}(x_s - x' + 1)$$

where x is the transformed value, x' the original value, and x_s either the maximum or minimum value.

Table 4.3 shows the influence of these transformations on a distribution which is uniformly distributed (the numbers 1 through 100). It is apparent that a skewed

Table 4.3   Effect of Logarithm and Square Root Transformations

Class number	Class bounds* Logarithm	Class bounds* Square root	Frequency distribution (%) Logarithm	Frequency distribution (%) Square root
	0.5	0.5		
1			1	5
	1.5	5.5		
2			2	7
	3.5	12.5		
3			4	11
	7.5	23.5		
4			6	14
	13.5	37.5		
5			13	18
	26.5	55.5		
6			25	20
	51.5	75.5		
7			49	25
	100.5	100.5		

* Assumes data range from 1 to 100. Logarithm and square root of 100 divided into 7 equal intervals then converted back to original data form. (Rounded to nearest 0.5.)

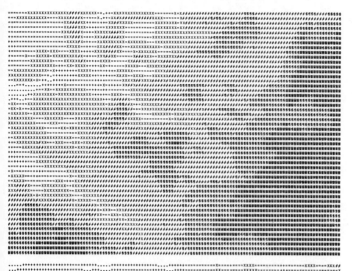

ORIGINAL

LOGARITHMIC

SQUARE ROOT

Figure 4.19 Line printer maps of the logarithmic and square root transformations.

distribution which has been transformed will scale into a relatively even distribution of map classes. Figure 4.19 illustrates the effect of these transformations with data.

Problems

(Some suggestions about solutions follow.)

1 Rewrite MAP1 so that it will accept more than 41 types by adding overprinting.
2 Modify MAP1 so that character locations on boundaries are set to blank.
3 Add coding to MAP2 so that it prints the area in each class (percentage of total) in the map legend.
4 Add coding to MAP1 or MAP2 which plots a frequency histogram of map areas.
5 Modify MAP1 or MAP2 so that line numbers and character numbers are printed along the left and top margins of the map. Character numbers are printed so that the digit is on one line, the tens position above it, and the hundreds above that, for example,

```
                                  11111...
        111111111122222...........99900000...
        123456789012345678901234...........78901234...
```

6 Rewrite MAP2 or MAP4 so that a specified value (such as 0.) is not included in the scaling calculations, and is set to blank on the output page.
7 Write a program which generalizes a quantitative map by two alternative methods, then maps each generalization and the difference between them with MAP4 or SQUEZ2 and MAP2. The two generalization techniques are:
   1 reduce an M by N map to a M/K by N/K map by sampling every Kth cell in both row and column directions;
   2 reduce an M by N map to an M/K by N/K map by taking the mean value of K by K blocks of cells.
8 Write a main program for MAP2 or MAP4 which converts a matrix of values to class numbers as specified by class intervals read as part of the data.
9 Rewrite MAP2A so that the class limits are estimated by sampling. Arguments to the subroutine will include the initial and final sampling densities, the increment between them, and the allowable tolerance, that is, the amount the worst class can deviate from the ideal. The modified MAP2A selects the sample, sends it to SORT for sorting, determines class limits, and tabulates and tests a frequency distribution using the entire data set. If the tolerance is not met, the sampling density is incremented and the process repeated either until the tolerance criterion is satisfied, or the upper sampling density limit is reached.
10 Write a program which reads a choropleth map coded by the length of traverse lines (as illustrated in figure 4.3), and converts this to matrix form. Assume each data card refers to a single portion of a traverse line, and contains the map row number, region value, and starting and ending column coordinates for that particular part of traverse line. Assume also that the total number of data cards is known. Check the result with SQUEZ2 and MAP2.

Suggestions for the Solution of Problems

1 Character combinations for overprinting should be checked carefully, or the results will be very monotonous and difficult to interpret. It is usually more

effective to use combinations which give a distinctive texture, rather than those which give a uniform grey tone.

An efficient way to organize the character set is as a fairly large number of print characters, and a much smaller set of characters which are used to over-print. If there are 20 print characters, the first 20 types are represented by the 20 characters overprinted by the first character in the overprint set. The second 20 types have the same print characters, but are overprinted by the second character in the overprint set, and so on.

2   This is most easily done by checking down rows and across columns for boundary changes, and setting one of the two possible character locations on either side of the boundary change to blank. For moving down columns, the following coding would be suitable:

```
        DO 900 J=1,NLINES
        DO 900 I=2,NCHARS
900     IF(MAP(I,J).NE.MAP(I,J-1).AND.MAP(I,J-1).NE.41) MAP(I,J)=41
```

There is a distortion introduced by this approach which may be substantial if regions are small.

3   The same procedure is followed as in MAP1.

4   For each type or class, use LINE to output a line of asterisks or similar symbols, followed by blanks. The number of asterisks is proportional to the frequency (area) of the type or class.

5   The WRITE and FORMAT for the character output are modified so that the DO variable which controls the WRITE can also be printed in it. Thus card 39 in MAP2 would become

```
        WRITE(6,31) J,(LINE(I,1),I=1,NCHARS)
```

The character numbers at the top of the page are produced by setting up a vector with values computed from the subscript of the element number. The tens line, for example, is given by $I/10$, and the units by $MOD(I, 10)$, where I is the character number. This vector is then written with the appropriate format.

6   Add an eighth level to TABLE which is blank. Keep elements with specified values out of scaling calculations with IF statements, and set to 8 in grey level assignment section.

7   K is the increment value in each DO statement.

8   Be certain that the entire range of the data values are specified as part of one class or another. If there is a gap, write an error message and set that value to a default class.

5

Irregularly Spaced Data with Explicit Coordinates

The matrix is the most convenient form for storing spatial data which are to be mapped with a line printer. The matrix has two major disadvantages for computer mapping, however, both involving the initial digitizing:

1 the procedure of superimposing a transparent grid on a source map is often so expensive in time and effort (whether done manually or by machine) that many computer studies are simply not feasible;

2 much spatial data does not exist as maps, but as lists of values and locations.

This chapter considers alternatives to the matrix grid approach for quantitative distributions in which irregularly spaced data points are coded by value and coordinates. These points may be the location of single events, individuals, or places (example 11), sample points on an isarithmic map (example 12), or centroids of regions (also example 12). No new FORTRAN statements are presented.

5.1 Coordinate Systems

The location component of spatial data may be explicitly described in one of three coordinate systems: spherical, polar, or rectangular. Spherical coordinates describe locations on the surface of the earth in terms of angular distance from the equator (latitude) and the Greenwich meridian (longitude). Example 5, ARCDIS, was concerned with measuring distances between points in this coordinate system. Polar coordinates describe the position of a point in terms of an angle and a distance from an origin (figure 5.1). They are rarely used with spatial data, although they could be very

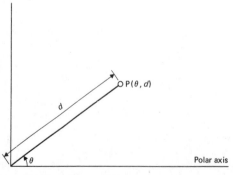

Figure 5.1 Polar coordinates.

useful for problems involving distributions around a central point where distances and relative orientations are important (such as small towns surrounding a metropolitan centre).

Rectangular coordinates are the usual method for describing the location of spatial data. In the classical Cartesian system, the coordinates of a point are expressed relative to an origin (0., 0.) at the lower left of the area of interest (figure 5.2). This system

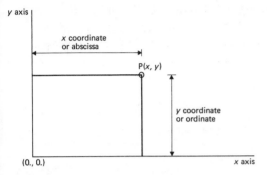

Figure 5.2   Cartesian coordinates.

may be generalized to include the other three quadrants around the origin, with the signs of the x coordinate (abscissa) and y coordinate (ordinate) changing accordingly (figure 5.3).

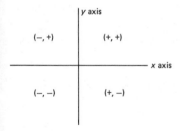

Figure 5.3   Signs of x and y coordinates by quadrant.

5.2   Transformation of Coordinates

It is apparent from chapter 4 that spatial data which are to be displayed with a line printer mapping system must be referenced in terms of rows and columns in the machine. This does not mean that irregularly spaced data points need to be coded in a special way for a computer, but rather that data distributions measured in a conventional Cartesian system must be translated in the machine to the form required for mapping. This is a relatively simple operation provided one is clear about row and column orientations on the map and in the machine.

For example, assume that X and Y are vectors of coordinates measured from an origin in the lower left of a map which have been read into the machine. These values must be converted to row and column form and stored in the vectors R and C. If the map is stored in transpose form (the assumption of all the examples in chapter 4), then the translations can be coded in FORTRAN as

```
      DO 10 K=1,NPTS
      R(K)=X(K)+1.
10    C(K)=YMAX-Y(K)+1.
```

where NPTS is the number of coordinate pairs, YMAX is the maximum Y coordinate value, and the measurement units in the two systems are equal (1 unit of X equals 1 unit of R, and 1 unit of Y equals 1 unit of C). It is assumed that the origin of the row and column system is 1., 1. rather than 0., 0. because row and column coordinates are commonly used as subscripts, and FORTRAN does not allow zero subscripts. Figure 5.4 illustrates the relationships involved in this transformation.

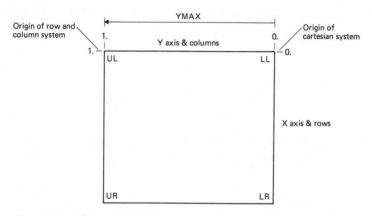

Figure 5.4   Relationship of Cartesian coordinates and row and column coordinates when the map is stored in the machine in transpose form. UL, LL, UR, and LR refer to the orientation of the source map (upper left corner, lower left corner, etc.).

If the map is considered to be stored in the machine with the same orientation as on the source, then the translation is given by

```
      DO 10 K=1,NPTS
      R(K)=YMAX-Y(K)+1.
10    C(K)=X(K)+1.
```

It is unusual that the units of X and Y are the same as those of R and C. As a result, translation is usually associated with scaling of the coordinates to fit the matrix area. This may be done for the transposed map matrix by

```
      FACTOR=RANGE/(SIZE-1)
      DO 10 K=1,NPTS
10    C(K)=(YMAX-Y(K))/FACTOR+1.
      R(K)=(X(K)-XMIN)/FACTOR+1.
```

and for the matrix with the same orientation as the source:

```
      FACTOR=RANGE/(SIZE-1)
      DO 10 K=1,NPTS
      R(K)=(YMAX-Y(K))/FACTOR+1.
10    C(K)=(X(K)-XMIN)/FACTOR+1.
```

where SIZE is the number of elements on the longest size of the matrix, and RANGE is the range of values on the corresponding coordinate axis.

A third transformation for rectangular coordinates which is required less often than translation or scaling is rotation. In mathematical terms this is

$$x = x' \cos \theta - y' \sin \theta$$
$$y = x' \sin \theta + y' \cos \theta$$

where x' and y' are the coordinates of the point before rotation, and θ is the angle of rotation (figure 5.5). The most common application of rotation in spatial problems is to re-compute coordinate values so that a scatter of points will fit into a smaller map area (figure 5.6). FORTRAN statements for this operation must be written with care.

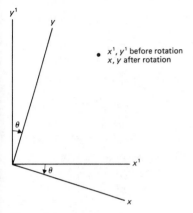

Figure 5.5   Rotation of axes.

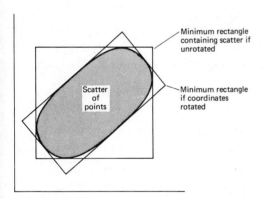

Figure 5.6   Change in required map area after rotation.

If the identities above are directly translated into assignment statements within a DO loop, i.e.,

```
DO 20 K=1,NPTS
    X(K)=X(K)*COS(THETA)-Y(K)*SIN(THETA)
20  Y(K)=X(K)*SIN(THETA)+Y(K)*COS(THETA)
```

then the resulting values of X will be correct, but all the values of Y will be incorrect because the X values used to compute them are already rotated. This coding is not only wrong, but inefficient as well, because the trigonometric functions are re-computed with each pass, when they could have been set before the loop started:

```
      U=COS(THETA)
      V=SIN(THETA)
      DO 20 K=1,NPTS
      TX=X(K)*U-Y(K)*V
      Y(K)=X(K)*V+Y(K)*U
20    X(K)=TX
```

Conversion from one system to another is quite simple for polar and rectangular coordinates since they are both planar. The relevant transformation equations are

$$x = d \cos \theta$$
$$y = d \sin \theta$$

Conversion between polar or rectangular and spherical coordinates however, usually involves quite complicated equations. Even a simple projection such as the Mercator requires the following calculations:

$$x = \{(u - u')/360\} \cdot 2\pi r$$
$$y = r \log_e \{\tan (45 + v/2)\}$$

where r is a scale factor (radius of the generating globe), u and v are latitude and longitude, and units are degrees. The computations required for more useful projections such as the universal transverse Mercator (UTM) can involve dozens of terms (over 60 for the UTM). These are not significant programming problems since they concern assignment statements only. However, they can involve significant computing costs since there are many trigonometric operations.

5.3   Example 11. SCALE and POINT: Scaling and Plotting Points with a Line Printer

Plotting the location of points is a basic operation when working with irregularly spaced data values. It would also seem to be a relatively simple task. Once the coordinate values were scaled by one of the techniques just discussed, they could be located in a matrix which had been previously initialized to a value representing blank with statements such as

```
      DO 1 K=1,NPTS
1     MAP(R(K),C(K))=1
```

where R and C are row and column coordinates. This matrix could then be mapped with MAP1.

This approach is adequate for the task, but is relatively wasteful of computer resources, particularly storage space. This example describes an alternative method which does not use a matrix, but a vector in which each line is stored, then printed. It is assumed that the coordinates are in Cartesian form. They are scaled to fit specified map dimensions by SCALE, sorted in ascending line order by SORT (the sorting routine presented in example 10), and printed by POINT as asterisks within a bordered area.

There are several ways in which these subroutines could be organized with a main program. For example, a chain of calls could link them together: MAIN calls SCALE, which calls SORT, which calls POINT. (This is the approach used in SQUEZ1, SQUEZ2, and MAP2A.) An alternative is to have MAIN call each of the three subroutines in turn, with the results of each computation being passed back to MAIN

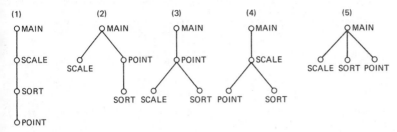

Figure 5.7   Alternative configurations for example 11.

before it is sent to the next subroutine. These and some other possibilities are illustrated in Figure 5.7.

All of these alternatives are equally efficient in terms of speed and require exactly the same number of FORTRAN statements. There are advantages to certain of these configurations, however, if any of the subroutines are to be used in other programs. A general rule is that any subroutine of general value should contain calls only to subprograms which always occur in association with it. For example, SQUEZ1 must occur with MAP1, and SQUEZ2 with MAP2. If calls are placed in subroutines where they are not required, then they must be removed or changed when that subroutine is used in another program. This is a minor inconvenience, but a common source of errors. If subroutines have been compiled from source into object form, these changes become considerably more costly.

In this example, SCALE and SORT are of general value, but POINT will probably be used only in this program. This means that configurations 2, 3, and 5 in figure 5.7 are preferable. Number 2 is the one used.

SCALE is a relatively elementary block of coding. In fact, only two assignment statements controlled by a DO statement are required to do the actual scaling calculations. The subroutine looks more complicated, however, because 17 statements are required to determine the values of the variables used in the scaling, the minimum Y coordinate, the maximum Y coordinate, and the scaling factor.

The coding starts with the usual block of comment cards and declarations. The same assumption is made as in earlier examples, that the program is operating on the transpose of the source map, that is, the origin of the coordinate system which was in the lower left of the original map, is in the upper right of the map in the machine. One option is provided in SCALE. If the argument MODE is equal to 1, then the coordinates are scaled to fit a six line per inch printer; POINT requires that the data be scaled in this way. (Further options are suggested in the problems at the end of this chapter.) Note that the coordinates are stored in matrix A which has two rows and NPTS columns, rather than as two separate vectors. This is done so that SORT may be used without revision when it is called later from POINT. Otherwise, either SORT would have to be rewritten to accept two vectors as arguments rather than a matrix, or the two vectors would have to be assigned to a matrix before SORT is called, and moved back to vector form on return.

```
      SUBROUTINE SCALE(A,NPTS,M,N,MODE)                          SCAL   1
C SCALES AND CONVERTS X AND Y COORDINATES INTO ROW AND COLUMN    SCAL   2
C COORDINATES WHICH CAN BE USED AS SUBSCRIPTS                    SCAL   3
C IN A MATRIX M ROWS BY N COLUMNS                                SCAL   4
C ON INPUT, ROW 1 OF A CONTAINS X COORDINATES, AND ROW 2         SCAL   5
C Y COORDINATES                                                  SCAL   6
```

```
C ROW COORDINATES ARE RETURNED IN ROW 1 OF A, AND COLUMN COORDINATES    SCAL    7
C IN ROW 2 OF A                                                         SCAL    8
C IT IS ASSUMED THAT THE X AXIS CORRESPONDS WITH ROWS                   SCAL    9
C (INCREASING AS X INCREASES), AND THAT THE Y AXIS CORRESPONDS          SCAL   10
C WITH COLUMNS (DECREASING AS Y INCREASES), THAT IS, THE                SCAL   11
C SOURCE MAP IS STORED IN TRANSPOSE FORM                                SCAL   12
C IF MODE=1, COLUMN COORDINATES ARE REDUCED TO FIT SIX LINE PER INCH    SCAL   13
C PRINTER.  N THEN REFERS TO NUMBER OF LINES IN OUTPUT MAP, AND M       SCAL   14
C THE NUMBER OF CHARACTERS                                              SCAL   15
      DIMENSION A(2,NPTS)                                               SCAL   16
C IF REQUIRED, CONVERT Y(COLUMN) COORDINATES                            SCAL   17
      IF(MODE.NE.1) GO TO 150                                           SCAL   18
      THRFIF=3./5.                                                      SCAL   19
      DO 160 K=1,NPTS                                                   SCAL   20
160   A(2,K)=A(2,K)*THRFIF                                              SCAL   21
150   CONTINUE                                                          SCAL   22
```

The main part of the program computes maximum and minimum X and Y values, ranges, and the value of the scaling factor.

```
C FIND MAXIMUM AND MINIMUM COORDINATE VALUES                           SCAL   23
C INITIALIZE                                                           SCAL   24
      XMAX=A(1,1)                                                       SCAL   25
      YMAX=A(2,1)                                                       SCAL   26
      XMIN=A(1,1)                                                       SCAL   27
      YMIN=A(2,1)                                                       SCAL   28
      DO 20 K=2,NPTS                                                    SCAL   29
      IF(A(1,K).GT.XMAX)XMAX=A(1,K)                                     SCAL   30
      IF(A(1,K).LT.XMIN)XMIN=A(1,K)                                     SCAL   31
      IF(A(2,K).GT.YMAX)YMAX=A(2,K)                                     SCAL   32
      IF(A(2,K).LT.YMIN)YMIN=A(2,K)                                     SCAL   33
20    CONTINUE                                                          SCAL   34
C CALCULATE RANGES                                                      SCAL   35
      XRANGE=XMAX-XMIN                                                  SCAL   36
      YRANGE=YMAX-YMIN                                                  SCAL   37
C CALCULATE RATIO OF EACH RANGE TO CORRESPONDING MATRIX DIMENSION       SCAL   38
      RX=XRANGE/(M-1)                                                   SCAL   39
      RY=YRANGE/(N-1)                                                   SCAL   40
C SCALING FACTOR IS LARGER OF TWO RATIOS                                SCAL   41
      FACTOR=AMAX1(RX,RY)                                               SCAL   42
```

The data are then scaled:

```
      DO 300 K=1,NPTS                                                   SCAL   43
      A(1,K)=(A(1,K)-XMIN)/FACTOR+1.                                    SCAL   44
300   A(2,K)=(YMAX-A(2,K))/FACTOR+1.                                    SCAL   45
      RETURN                                                            SCAL   46
      END                                                               SCAL   47
```

On return from SCALE, the coordinates are in a form which could be used directly as subscripts to designate character locations in an output matrix. They are then sent to POINT. Most of this subroutine is concerned with printing horizontal and vertical lines defining the borders of the map area. In the initial section, for example, symbols are defined and a horizontal line made up of dashes is printed across the top of the map area:

```
      SUBROUTINE POINT (WHERE,NPTS,NCHARS,NLINES)                       PT      1
C PRINTS * IN CHARACTER LOCATION SPECIFIED BY WHERE                     PT      2
C ROW 1 OF WHERE IS CHARACTER NUMBER ON LINE                            PT      3
C AND ROW 2 IS LINE NUMBER                                              PT      4
      DIMENSION WHERE(2,NPTS)                                           PT      5
C NCHARS IS NUMBER OF CHARACTERS PER LINE (LESS THAN 129 TO ALLOW       PT      6
C FOR MARGINS OF MAP)                                                   PT      7
C NLINES IS TOTAL NUMBER OF LINES (NO LIMIT)                            PT      8
C LINE IS USED TO HOLD ONE LINE FOR OUTPUT                              PT      9
      REAL LINE(131)                                                    PT     10
      DATA BLANK,AST,HOR,VER/' ','*','-','|'/                           PT     11
C NCHARS+1 AND NCHARS+2 ARE REQUIRED IN DO LOOPS                        PT     12
      NCHAR1=NCHARS+1                                                   PT     13
      NCHAR2=NCHARS+2                                                   PT     14
```

```
C PRINT LINE ACROSS TOP OF MAP                              PT    15
      DO 50 I=1,NCHAR2                                      PT    16
50    LINE(I)=HOR                                           PT    17
      WRITE(6,51) (LINE(I),I=1,NCHAR2)                      PT    18
51    FORMAT('1',131A1)                                     PT    19
      LINE(NCHAR1)=VER                                      PT    20
```

The coordinates are then sorted so that they are arranged with increasing line numbers:

```
C SORT COORDINATES SO THAT LINE VALUES INCREASE            PT    21
      CALL SORT(WHERE,NPTS,2,2)                             PT    22
```

The main output loop is executed as many times as there are lines to be printed. Each time, the output vector LINE is initialized to blank, all character locations coinciding with a point set to an asterisk, and LINE printed. The trick in this coding is to allow for lines with no points, with a single point, and with multiple points.

```
C START OUTPUT LOOP                                        PT    23
      K=1                                                  PT    24
      DO 200 J=1,NLINES                                    PT    25
C SET LINE TO BLANK                                        PT    26
      DO 201 I=1,NCHARS                                    PT    27
201   LINE(I)=BLANK                                        PT    28
203   IF(K.GT.NPTS)GO TO 200                               PT    29
      IF(IFIX(WHERE(2,K)).NE.J) GO TO 200                  PT    30
      LINE(WHERE(1,K))=AST                                 PT    31
      K=K+1                                                PT    32
      GO TO 203                                            PT    33
200   WRITE(6,210) (LINE(I),I=1,NCHAR1)                    PT    34
210   FORMAT(' |',130A1)                                   PT    35
```

Note that the boundary line on the left side is set by the FORMAT statement, while the one on the right side is stored in LINE by an assignment statement before the loop (card PT 20).

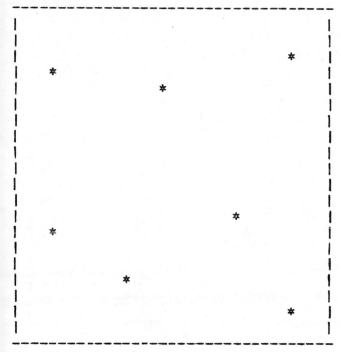

Figure 5.8   Some output from example 11.

The last part of the subroutine prints a line across the bottom of the map, and skips to a new page:

```
C PRINT LINE ACROSS BOTTOM OF MAP                                    PT   36
      DO 52 I=1,NCHAR2                                               PT   37
52    LINE(I)=HOR                                                    PT   38
      WRITE(6,53)(LINE(I),I=1,NCHAR2)                                PT   39
53    FORMAT(' ',131A1,/,'1')                                        PT   40
      RETURN                                                         PT   41
      END                                                            PT   42
```

Writing a main program to read the data and call SCALE and POINT will be left as an exercise for the reader.

5.4 Isarithmic Maps from Irregularly Spaced Data Points

One of the most common and useful applications of computers to spatial data is the computation of line printer maps from a set of irregularly spaced sample points in which isarithms (lines joining points of equal value) are defined by the border between grey level zones (figure 5.9). The sample points are sometimes measured from an existing source map, but are more often obtained from lists of spatial distributions (coordinates and values). In fact, even if a map already exists, it is better to use the data values from which the map was originally compiled, rather than take samples from the finished product. This is because the computer is able to proceed from data values to map directly. Measuring from the map adds the interpretation and bias introduced by the map compiler, as well as measurement error; we are coding the source data as filtered by the map.

It is important to recognize that there are two kinds of isarithmic maps, differing in the location accuracy of the original data points. In the first, the control points are precisely located (such as weather stations which are used for maps of isotherms or isohyets). Isarithms derived from such data points are termed isometric lines (Hsu and Robinson, 1970). In the second kind of isarithmic map, the data points are the centroids of regions such as census tracts or municipalities, and therefore much less precisely located. Isarithms of this second type are termed isopleths by Hsu and Robinson. The positioning of the centroid points has a significant effect on the resulting isopleth map. A common problem arises when these points are located in the estimated centre of the *area*, but are used to map distributions which are unevenly distributed within each area.

If an isarithmic map must be digitized because the source data are unavailable, there are many possible ways to position a set of irregularly spaced sample points. The most appropriate system in the statistical sense is one in which points are located randomly on the map, and a sufficient number of points are used that such measures as the mean and standard deviation of the map are estimated with confidence. An alternative which would require fewer sample points but not meet the assumptions of statistical analysis would be to position the points at specific important locations on the map such as the tops of hills, the edges of plains, along valley floors and ridges, and at changes in slope. This technique has the disadvantages that:

1   the point locations very much reflect an individual's conception of what is important in the map;
2   there is a natural tendency to omit small features; and

Figure 5.9   An Isarithmic map produced by a line printer.

3   there is no guarantee that the points adequately capture all the variation of interest in the problem.

There are two techniques within these extremes that the author has found useful: a partially random sample using points, and a systematic sample using a grid. The first of these is described in technical terms as a stratified systematic unaligned sample (Berry and Baker, 1968). It involves first selecting a grid (taking into account the same

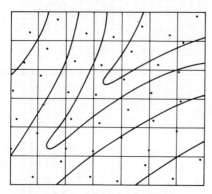

Figure 5.10   A stratified systematic unaligned sample.

kind of considerations as were described in chapter 4), and then positioning sample points within each grid square using the following procedure (illustrated in figure 5.10):

1   select two random numbers (usually from a random number table) and use these as the horizontal and vertical coordinates of a point within the grid cell in the upper left corner;
2   position points within each grid cell in the first column of the grid using the same vertical coordinate as in the first grid cell for all other cells, but selecting the horizontal coordinate for each randomly;
3   position points within each grid cell in the first row of the grid using the same horizontal coordinate as in the first grid cell for all other cells, but selecting the vertical coordinate for each randomly;
4   position points within the remaining grid cells using the horizontal coordinate of the point in the first column of each cell's row, and the vertical coordinate of the point in the first row of each cell's column;
5   interpolate the value at each point by inspection of surrounding isarithms.

The stratified systematic unaligned sample is not only considerably easier to apply than purely random techniques, but is also more accurate than random, stratified random, or even systematic procedures for most spatial distributions. It is particularly useful when there are any periodicities in the data (such as a set of parallel ridges).

A second procedure which is useful for digitizing isarithmic maps consists simply of superimposing a grid on the map, and coding coordinates and values where a grid line intersects with an isarithm (figure 5.11). This is a systematic procedure which is rapid and ensures a higher sampling rate where there is more spatial variation, i.e., where there are more isarithms or they are more irregular. The accuracy of the procedure depends, of course, on the grid size and its orientation relative to the source map.

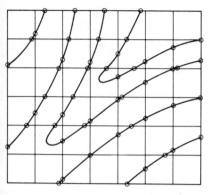

Figure 5.11   Systematic sampling of the intersections of isarithms with a superimposed grid.

5.5   Two-Dimensional Interpolation

A basic operation applied to sets of irregularly spaced data values is two-dimensional interpolation in order to estimate additional values. Interpolation is used in this chapter to fill values in a matrix which can then be mapped with a line printer using the techniques described in chapter 4.

The methods available for two-dimensional interpolation fall into two general categories: surface fitting and numerical approximation. In surface fitting, a polynomial equation is determined which best fits the pattern of data values, and interpolation takes place by substituting the coordinates of other points on the map and solving the equation. The procedure by which such equations are fitted is beyond the scope of this book. It involves the selection of a type and order of polynomial, and the estimation of its coefficients using the principle of least squares.

The surface described by this equation will rarely pass through the original data values, but will deviate above and below them. This is desirable for data which are 'noisy' and have a known type of spatial relationship (such as geological or geophysical). The surface 'cleans' the raw data in these cases. For most distributions, however, the magnitude of both error and spatial autocorrelation is unknown. This means that in cases where the data values are exact and the fit of the surface is poor, interpolation by surface fitting may produce quite misleading results.

Interpolation by numerical approximation assumes that the value at a given location on a map is determined by the distance to and the value of a number of surrounding data points. Compared with surface fitting, numerical approximation allows representation of more complex surfaces (particularly those with anomalous features), restricts the spatial influence of any errors, and puts the interpolated surface through the data points. It is the more appropriate method for most spatial data.

The simplest form of interpolation by numerical approximation assumes that the value at a specified map location is equal to the average of the values at a number of surrounding data points, weighted by the reciprocal of distance to those points. Thus

$$X = \sum_{k=1}^{M}(Z_k/D_k)\Big/\sum_{k=1}^{M}(1/D_k)$$

where X is the interpolated value, Z_k are the surrounding M data values, and D_k are the distances from X to these data points.

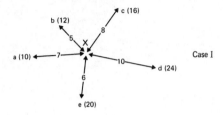

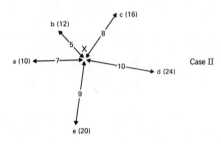

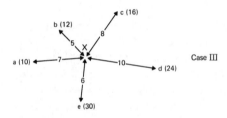

Figure 5.12   Three arrangements of data points around a location *X*.

Those unfamiliar with weighted averages may understand the principle better by considering figure 5.12 and table 5.1, which illustrate the effect of changing the value and location of a single sample point on the interpolation result.

It is possible to produce quite different maps using the basic weighted average interpolation formula simply by varying the number of data points used in the calculation. It is difficult to state a value for *M* which produces the best map because this will depend on the spacing of data points and the spatial variation within the map. However, one can state that if more than 12 points are used in each calculation, the resulting map will be highly smoothed, while less than four points will produce abrupt changes over the surface. The appropriate value seems to be in the range 6 to 9.

There are two elaborations on the basic weighted average scheme which are widely used because they appear to result in a more accurate map. A very common procedure is to use the square of distance in the formula, that is,

$$X = \sum_{k=1}^{M} (Z_k/D_k^2) \Big/ \sum_{k=1}^{M} (1/D_k^2)$$

This tends to put more weight on nearby points. When the values in table 5.1 are recomputed using the square of distance, the results are 15.2 for case I, 14.5 for case II, and 17.7 for case III, all smaller than previously, and largely because of the reduced influence of point D, farthest from X and with the highest value (except in case III).

Table 5.1   Weighted average calculations (for figure 5.12).

Sample point	Value at sample point (Z)	Distance from X to sample point (D)	Z/D	$1/D$
a	10	7	1.43	0.13
b	12	5	2.40	0.20
c	16	8	2.00	0.12
d	24	10	2.40	0.10
e (case I)	20	6	3.33	0.17
e (case II; value unchanged from I, but distance increased 50%)				
	20	9	2.22	0.11
e (case III; distance unchanged from I, but value increased 50%)				
	30	6	5.00	0.17

Case I:
$\sum Z/D = 11.56$
$\sum 1/D = 0.69$
Interpolated value $= 16.0$
Case II:
$\sum Z/D = 10.41$
$\sum 1/D = 0.66$
Interpolated value $= 15.8$
Case III:
$Z/D = 13.23$
$1/D = 0.71$
Interpolated value $= 18.6$

Tobler (1970) presents a version of the weighted average formula which empha-sizes local influences even further:

$$X = 0.5 \left\{ Z_n + \left(\sum_{k=1}^{M} (Z_k/D_k^2) \bigg/ \sum_{k=1}^{M} (1/D_k^2) \right) \right\}$$

where Z_n is the value of the nearest data point. When this formula is used with the data in table 5.1, the results are 14.0 for case I, 13.9 for case II, and 15.3 for case III. The effect of the relatively low value of point B is apparent.

5.6   Example 12. INTERP: Line Printer Maps from Irregularly Spaced Data

INTERP is a subroutine which fills a matrix with values interpolated from a set of irregularly spaced data points by numerical approximation (weighted averages). The output from the subroutine is a matrix in which each element represents a square or rectangular grid cell. depending on what was specified in the input arguments. In the former case, the matrix can be sent directly to MAP4 for mapping, and in the latter to MAP2.

The main program which calls INTERP must first read the coordinates and values of the data points and, if they are not scaled or in row and column form, call SCALE. (The final argument to SCALE must be set to 0 so that there is no reduction of column coordinates to line numbers.) The program then calls INTERP with arguments identifying the vectors of row and column coordinates and values, the matrix size, the number of data points to be used in the interpolation calculations, and an argument set to 0 or 1 to indicate a square or rectangular grid. If elements representing squares are desired, the argument is set to 0, and the matrix size in the arguments for INTERP is the same as that for SCALE. If a rectangular grid is desired, the argument is 1, and the number of columns specified for INTERP must be three-fifths of the number for SCALE. The final part of the main program will call MAP2 or MAP4. Writing such a main program is left as an exercise for the reader.

The initial section of INTERP specifies the variables used and describes their purpose:

```
      SUBROUTINE INTERP(R,C,Z,NPTS,MAP,NROWS,NCOLS,M,MODE)          ITRP   1
C TWO DIMENSIONAL INTERPOLATION BY WEIGHTED AVERAGE OF NEAREST M POINTS ITRP   2
C SUBROUTINE REQUIRED: SORT                                         ITRP   3
C POINT COORDINATES ARE STORED IN R (ROW), C (COLUMN)               ITRP   4
C ROW COORDINATES .GE. 1 .AND. .LE.NROWS                            ITRP   5
C COLUMN COORDINATES .GE. 1 .AND. .LE.NCOLS                         ITRP   6
C NPTS IS NUMBER OF POINTS                                          ITRP   7
      DIMENSION R(NPTS),C(NPTS),Z(NPTS)                            ITRP   8
C MAP IS THE OUTPUT MATRIX, NROWS BY NCOLS                          ITRP   9
      REAL MAP(NROWS,NCOLS)                                        ITRP  10
C IF MODE=1, MAP REPRESENTS A RECTANGULAR GRID FOR COMPUTER MAPPING ITRP  11
C WITH A SIX LINE PER INCH PRINTER                                  ITRP  12
C M IS THE NUMBER OF DATA POINTS TO BE USED IN THE INTERPOLATION    ITRP  13
C DIST CONTAINS DISTANCES FROM THE CENTER OF A GRID CELL TO ALL     ITRP  14
C DATA POINTS (ASSUME MAXIMUM OF 1000)                              ITRP  15
      DIMENSION DIST(1000)                                         ITRP  16
C WITHIN CONTAINS DISTANCES TO (ROW 1) AND VALUES OF (ROW 2)        ITRP  17
C DATA POINTS WITHIN SEARCH RADIUS                                  ITRP  18
      DIMENSION WITHIN(2,1000)                                     ITRP  19
```

The interpolation computations for each matrix element consists of three steps:

1 compute the distances between the centre of the grid cell and all data points;
2 determine the *M* smallest distances (using SORT);
3 solve the interpolation equation

$$X_{ij} = \sum_{k=1}^{M} (Z_k/D_k{}^2) \bigg/ \sum_{k=1}^{M} (1/D_k{}^2)$$

and store the result in the matrix element.

The FORTRAN coding for these three steps is presented below under the control of two DO statements. Also incorporated in these statements is the allowance for either square or rectangular cells (cards 24, 25, and 31), a check for cases where a data point coincides with the centre of a cell (cards 36, 39, and 40), and a technique which reduces the time required in the second step. This consists of a search radius set to an initial value before the loop (card 23), and used to limit the size of arrays sent to SORT (card 44). If this radius is too small, it is reset to a value of 25 per cent larger (card 51), and if it is too large, it is reduced by 10 per cent (card 65).

```
C RADIUS DEFINES INITIAL SEARCH AREA                               ITRP  20
C SET INITIALLY TO TWICE VALUE REQUIRED TO CONTAIN M POINTS IF POINTS ITRP  21
C ARE EVENLY DISTRIBUTED                                           ITRP  22
      RADIUS=SQRT(2.*(NROWS*NCOLS)/NPTS)                           ITRP  23
      SCALER=1.                                                    ITRP  24
      IF(MODE.EQ.1) SCALER=5./3.                                   ITRP  25
```

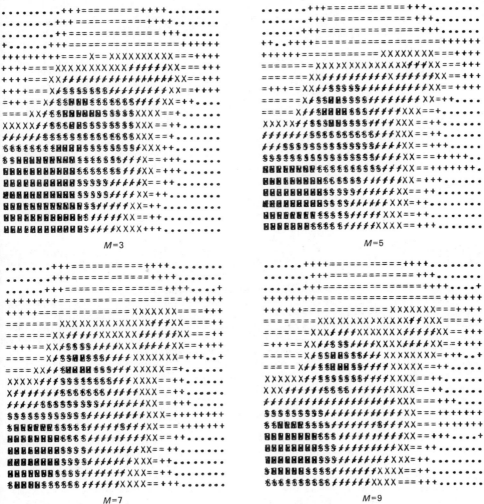

Figure 5.13 Line printer maps produced by INTERP and MAP2 using different values of *M*.

```
C START LOOP TO LOOK AT ALL MAP POINTS                           ITRP   26
      DO 3 J=1,NCOLS                                             ITRP   27
      DO 3 I=1,NROWS                                             ITRP   28
C CONVERT ROW AND COLUMN DO VARIABLES TO REAL FORM               ITRP   29
      ROW=I+0.5                                                  ITRP   30
      COL=(J+0.5)*SCALER                                         ITRP   31
C COMPUTE DISTANCES FROM CENTER OF THIS CELL TO ALL DATA POINTS  ITRP   32
      DO 4 K=1,NPTS                                              ITRP   33
C USE MULTIPLICATION INSTEAD OF SQUARING TO SPEED EXECUTION      ITRP   34
      DIST(K)=SQRT((ROW-R(K))*(ROW-R(K))+(COL-C(K))*(COL-C(K)))  ITRP   35
4     IF(DIST(K).EQ.0.) GO TO 41                                 ITRP   36
      GO TO 45                                                   ITRP   37
C DATA POINT COINCIDES WITH CENTER OF THE GRID CELL              ITRP   38
41    MAP(I,J)=Z(K)                                              ITRP   39
      GO TO 3                                                    ITRP   40
C FIND ALL POINTS WITHIN RADIUS                                  ITRP   41
45    MM=0                                                       ITRP   42
      DO 40 K=1,NPTS                                             ITRP   43
      IF(DIST(K).GT.RADIUS) GO TO 40                             ITRP   44
      MM=MM+1                                                    ITRP   45
      WITHIN(1,MM)=DIST(K)                                       ITRP   46
      WITHIN(2,MM)=Z(K)                                          ITRP   47
```

```
40      CONTINUE                                                      ITRP  48
C IF INSUFFICIENT POINTS, INCREASE RADIUS BY 25% AND REDO LOOP       ITRP  49
        IF(MM.GT.M) GO TO 50                                         ITRP  50
        RADIUS=RADIUS*1.25                                           ITRP  51
        GO TO 45                                                     ITRP  52
C FIND M SMALLEST DISTANCES                                          ITRP  53
C SORT BY DISTANCE TO FIND M NEAREST POINTS                          ITRP  54
50      CALL SORT(WITHIN,MM,2,1)                                     ITRP  55
C SOLVE INTERPOLATION FUNCTION                                       ITRP  56
        SUM1=0.                                                      ITRP  57
        SUM2=0.                                                      ITRP  58
        DO 7 K=1,M                                                   ITRP  59
        SQUARE=WITHIN(1,K)*WITHIN(1,K)                               ITRP  60
        SUM1=SUM1+WITHIN(2,K)/SQUARE                                 ITRP  61
7       SUM2=SUM2+1./SQUARE                                          ITRP  62
        MAP(I,J)=SUM1/SUM2                                           ITRP  63
C REDUCE RADIUS IF TOO MANY POINTS ARE FALLING WITHIN SEARCH RADIUS  ITRP  64
        IF(MM.GT.2*M) RADIUS=RADIUS*.9                               ITRP  65
    3 CONTINUE                                                       ITRP  66
        RETURN                                                       ITRP  67
        END                                                          ITRP  68
```

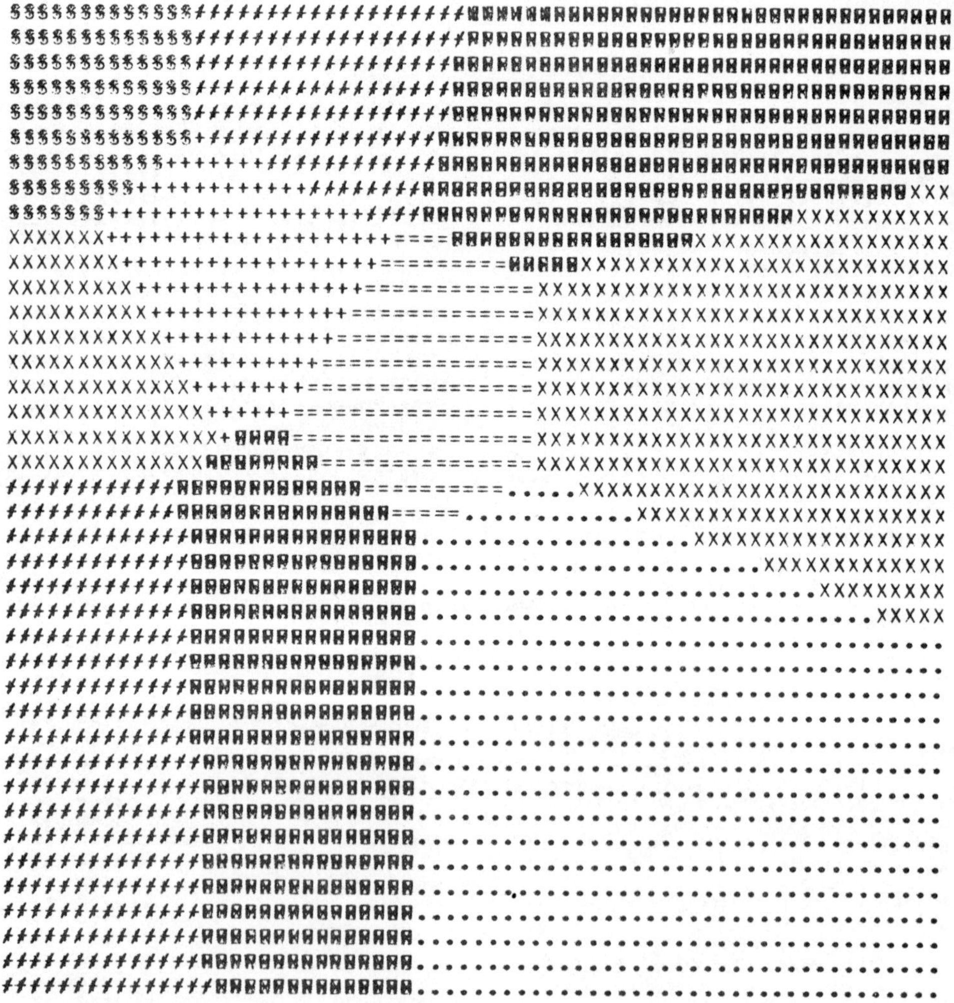

Figure 5.14 A proximal map.

5.7 Choropleth Maps from Irregularly Spaced Data Points (Proximal Maps)

If the pattern is not too complex, it is possible to generate line printer choropleth maps from centroid information (coordinates and value). This is done using INTERP with M (the number of data points used in the interpolation) set to 1. Thus

$$X_{ij} = (Z_n / D_n^2)/(1/D_n^2)$$

which simplifies to

$$X_{ij} = Z_n$$

where Z_n is the value of the nearest data point. The result is a map in which the value of each character location is that of the nearest data point (sometimes called a *proximal* map).

This technique is obviously a simple and quick way to produce choropleth maps, particularly when compared with boundary encoding procedures which will be described in the following chapter. However, there is often a substantial loss of accuracy in the mapping process. In fact, the only choropleth map which can be precisely rendered by a line printer map using INTERP is one in which all regions are of equal size and their centroids are equidistant, that is, a pattern of hexagons.

Either quantitative or qualitative (numerical label) data can be mapped in this way, provided one is careful to convert the qualitative data to real form before sending it to INTERP, and change it back to integer form before called MAP1.

Problems

(Some suggestions about solutions follow.)
1 Draw a flowchart for POINT.
2 Draw a flowchart for INTERP.
3 Write a program which reads the coordinates of a set of points and computes:
    1 the centre of the point distribution (the location with mean x and y or mean row and column coordinates); and
    2 the standard deviation of distances from all points to the centre of the distribution (also called the standard distance).
4 Rewrite SCALE and POINT to provide one or more of the following options:
    1 label points with numbers or names;
    2 label the left and upper borders with coordinate values;
    3 centre the point pattern within the specified matrix dimensions.
5 Rewrite INTERP so that matrix cells which contain a data value are set to that value, and interpolation takes place only for those elements which contain no data value. This version of INTERP is considerably faster than that presented in the text, but with a possible cost in reduced precision in those cells containing data values, either because they are well away from the centre of the cell, or because there are two or more in the same cell.
6 Modify INTERP and MAP2 so that a blank or special symbol is placed on the printer map at those locations where a data point occurs. (This is more easily done with the modified version described in problem 5 than the version presented in the text.)

7 Write a program which reads a set of irregularly spaced data values, then calls INTERP with values of M from 3 to 12, accumulating in a map matrix the difference at each cell from that previously computed in INTERP. The result, when mapped with MAP2, indicates those parts of the map which change in value a great deal as additional points are taken into account. This map may be used either to determine the best value of M to use for a larger data set of which these data were a part, or to locate areas where additional data points may be useful.

8 Assume that potential interaction (such as visitors or retail sales) can be estimated at a location by an equation of the form

$$I = \sum (P_k C_k / D_k^A) / \sum (1/D_k^A)$$

where I is interaction, P population in region k, C the propensity of individuals in region to interact, D the distance from the centre of the region to the point, and A a constant describing the friction of distance. Write a program which reads values for P and C as two gridded maps (matrices), and computes values for I for any specified coordinate locations.

Suggestions for the Solution of Problems

3 Example 2, MSDMAP, can be modified to do this.

4 The first and second modifications are done in POINT, and the third in SCALE.

 1 Labels are read in the main program with coordinates (using A1 format) and sent in the argument list to POINT. The label is included in the x output line as an assignment statement after card 31. If the label is more than one character long, it must be stored in an array, and several assignment statements or a DO loop is required in POINT.

 2 This is done in the same way as problem 5, chapter 4.

 3 One half the difference between the length of the two sides is added to the coordinates in the shorter dimension.

6 This may be done in several ways. The easiest is to set each element in the map matrix in which a data point falls to a special value (such as 999.), which is then detected in a revised MAP2 as each line is printed. Any matrix elements with this special value are then set to a special symbol. An alternative and somewhat faster technique is to sort the coordinate values as in POINT, and check this sorted list in MAP2 as each line is printed.

6

Lines and Networks

Spatial problems which involve lines, be they the boundaries of regions, networks, or mapping symbols, require procedures for coding, processing, and display which are quite different from those presented thus far. Lines on maps are digitized by the coordinates of end points of line segments, and networks are coded for analysis as a matrix describing the relationship among nodes. Display is possible with a line printer only after considerable processing, but plotters which are available at many computer centres present line information with considerable precision and minimal programming requirements.

This chapter describes major techniques by which line and network data are used in a computer environment. No new FORTRAN is presented, but three examples illustrate the use of a line printer with boundary data (example 13), the use of a line plotter (example 14), and network analysis (example 15).

6.1 Regions as Polygons: Digitizing by Boundary Coordinates

Two techniques have been presented in this book for digitizing regions from choropleth maps. The first of these, described in chapter 4, involves superimposing a grid and determining values in a matrix. This gives a reasonable accurate representation (depending on the fineness of the mesh), but is often very time-consuming and expensive. The second technique, described in chapter 5, references each region by the coordinates of a centroid, and the original map is reconstructed as a proximal map. The principal problem with this method is that it often results in substantial distortion. A third technique which is described here can be both accurate and reasonably quick, and has some computational advantages. It consists of digitizing the coordinates of selected points along the boundary of a region so that it can be represented as a polygon.

The selection of these boundary points is not a trivial problem. In fact, it is a central research topic in picture processing because of its importance in algorithms which form polygonal regions from grey level samples of digitized images. It is also of considerable importance in processing map boundaries which have been digitized by machines (scanners or coordinate digitizers) into more compact arrays. However, the user concerned with digitizing a few maps by hand need not concern himself with the literature on this problem. The additional cost of his inefficiency is inconsequential when compared with the cost of determining optimal sampling patterns.

This is not advice to position boundary points at will. In order to ensure that major variations in the shape of a region are captured with a relatively small number

After steps 1 and 2 (I=inflection point; C=centre of curve):

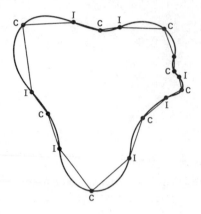

After steps 3 and 4 (A=points added in step 3; N=point added in step 4):

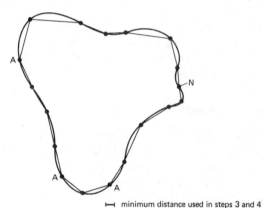

⊢⊣ minimum distance used in steps 3 and 4

Figure 6.1   Positions of boundary points at various stages in encoding.

of points, it is useful to establish specific procedures with defined criteria. The following is an example (illustrated in figure 6.1):

1   mark all inflection points (changes from concave to convex curves, vice versa, or from a straight line to a curve) and centres of curves; (curves which taper may require more than one point);
2   join (with pencil) adjacent points with straight lines; (with practice, this step can be eliminated);
3   if the deviation of the boundary from any straight line segment is greater than a specified minimum, add a point on the boundary where the deviation is greatest, and repeat steps 2 and 3;
4   replace any pair of boundary points which are within a specified minimum distance of one another by a single point between them, and repeat steps 2 and 3.

   The minimum distance referred to in steps 3 and 4 is the equivalent of the grid size for maps coded as regularly spaced values, and its determination involves similar considerations: it should be small enough that variation important to the problem is captured, but not so small that time is spent coding the artistic licence of the cartographer. There is little problem with this decision for boundaries of political units because most such boundaries are straight lines, and in those which are not, small irregularities are of no importance. In maps of natural regions, however, small variations in boundaries can be quite significant. Considerable care is required when dealing with such patterns.

Manual digitizing of an entire choropleth map is done by identifying and marking boundary points for all regions using the technique described above, labelling each with its coordinates (using a grid with a light table as an aid), and transferring these data to coding sheets and punch cards. The data are organized on the cards to indicate for each region its label (or value), the number of points in its boundary, then the coordinates for the boundary.

An alternative method which is useful in certain situations is to code the coordinates of each boundary point together with an arbitrarily assigned identifying number, and separately code for each region its label or value and the identifying numbers of the points which make up its boundary. The actual coordinate values can then be assigned to each region by a program in the machine. This technique is most useful when boundary points tend to be shared by several regions. A version of this technique is used in example 13.

Machines are often used to assist in the digitizing process. The most commonly available devices for this are graphic or coordinate digitizers. These consist of a flat surface on which the map is fastened, a cursor or similar device which is positioned on the location to be digitized, and a keypunch or tape machine on which the coordinate information is placed. The operator is able to enter an identifying label either with each coordinate pair or at the head of a string of coordinates. These machines can thus be used with either of the two techniques described above.

Some coordinate digitizers operate in a mode in which coordinates are measured and recorded at equal intervals along a line, rather than at designated points. These data are then thinned in a computer to the smallest set which adequately describes the region's shape, an interesting computing problem.

6.2   Example 13. BDRY2: Filling a Matrix from Boundary Coordinates

This example is a main program and a subroutine which read quantitative data using the second of the two techniques described above and produce a matrix where each element represents either a rectangular or square cell. This matrix may be used for computer mapping with MAP2 or MAP4, or for further computations such as calculating the area or perimeter of each type, or tabulating values from other maps which occur within each region.

The main program illustrates approaches to data reading which are often used when the data consist of several parts, each of any possible length (up to a maximum). In this case, the first part of the data consists of the x and y coordinates of boundary points or vertices, and the second part the numerical value of each region and the numbers of those boundary points which define it. These parts are separated for this program by what is termed a fence card, that is, a card with the same format as those preceding it, but with a value punched on it which is highly unlikely to occur in the data (such as 99999.). The program checks each card as it is read with an IF statement and when the fence card value is found, proceeds to the next part of the program.

The second part of the data is partitioned into sections referring to each region by leading each with a card on which is punched the region's value and the total number of points which define it. The program reads only the specified number of boundary point numbers, then looks for information on the next region. This part of the data deck is also terminated with a fence card. Figure 6.2 is an example of a short data

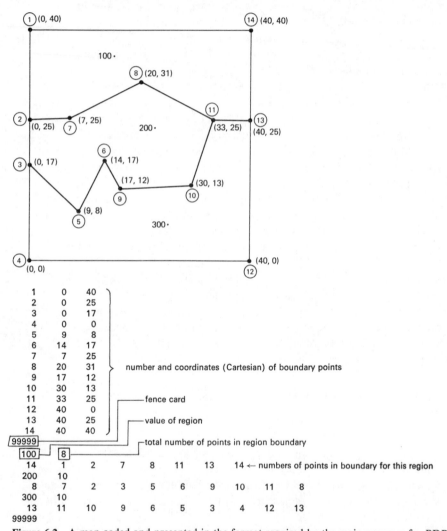

Figure 6.2   A map coded and presented in the format required by the main program for BDRY2.

deck arranged in this way. Note that the coordinates are Cartesian (they are converted to rows and columns by SCALE), and that the boundary points are arranged in the second part of the data with a specific order, that is, starting at the top point (or the top right point, if there is a tie), and proceeding counterclockwise. The reason for this order will become apparent shortly.

The main program presented below is written to produce a computer map up to 41 characters wide and 41 lines deep. The computations and printing for this are done by SCALE, BDRY2, POINT, and MAP2. The main program is primarily concerned with reading data and checking for fence cards. The general procedure for this is shown in figure 6.3. In this particular program, the procedures for counting are somewhat more complicated.

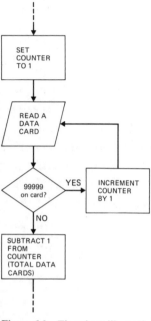

Figure 6.3 Flowchart illustrating use of fence cards.

The program starts with comments and specifications, then reads the coordinates of boundary points and, as a check, prints them:

```
C MAIN FOR BDRY2                                                     MAIN    1
C VERT IS ARRAY OF BOUNDARY COORDINATES                              MAIN    2
C XY IS ARRAY OF BOUNDARY COORDINATES ARRANGED BY REGION             MAIN    3
C NP IS ARRAY OF NUMBER OF POINTS IN EACH BOUNDARY                   MAIN    4
C (NEEDED TO LOCATE A REGION'S COORDINATES IN XY)                    MAIN    5
C Z IS ARRAY OF VALUES OF REGIONS                                    MAIN    6
C MAP IS OUTPUT MATRIX                                               MAIN    7
      REAL VERT(2,25),XY(2,100),NP(25),Z(25),MAP(41,41)             MAIN    8
      NCHARS=50                                                      MAIN    9
      NLINES=30                                                      MAIN   10
C READ X AND Y COORDINATES OF VERTICES UNTIL A FENCE CARD            MAIN   11
C ENCOUNTERED                                                        MAIN   12
      NVERT=1                                                        MAIN   13
12    READ(5,10)L,COORD1,COORD2                                      MAIN   14
10    FORMAT(I5,2F5.0)                                               MAIN   15
      IF(L.EQ.99999) GO TO 11                                        MAIN   16
      VERT(1,L)=COORD1                                               MAIN   17
      VERT(2,L)=COORD2                                               MAIN   18
      NVERT=NVERT+1                                                  MAIN   19
      GO TO 12                                                       MAIN   20
11    NVERT=NVERT-1                                                  MAIN   21
      WRITE(6,13)                                                    MAIN   22
```

```
13      FORMAT('1COORDINATES OF VERTICES READ:')              MAIN   23
        WRITE(6,20)(K,(VERT(J,K),J=1,2),K=1,NVERT)            MAIN   24
20      FORMAT('0',I5,2F10.2)                                 MAIN   25
```

In the part of the program which reads the second part of the data, the array XY has two purposes. In the statement on card 35 part of the first row of XY is used to store the numbers of the coordinates defining the NRth region. In the loop ending with statement 26, parts of both rows of XY are then set to the actual coordinates corresponding to the numbers stored earlier.

```
C REAC VALUE OF REGION AND NUMBER OF PCINTS IN BOUNDARY      MAIN   26
        NR=1                                                  MAIN   27
        KB=0                                                  MAIN   28
24      READ(5,30) Z(NR),NP(NR)                               MAIN   29
30      FORMAT(2F5.0)                                         MAIN   30
        IF(Z(NR).EQ.99999.) GO TO 22                          MAIN   31
        KA=KB+1                                               MAIN   32
        KB=KA+NP(NR)-1                                        MAIN   33
C READ NUMBERS OF BOUNDARY POINTS                            MAIN   34
        READ(5,23)(XY(1,I),I=KA,KB)                           MAIN   35
23      FORMAT(14F5.0)                                        MAIN   36
C STORE BOUNDARY COORDINATES IN XY                           MAIN   37
        DO 26 J=KA,KB                                         MAIN   38
        XY(2,J)=VERT(2,XY(1,J))                               MAIN   39
26      XY(1,J)=VERT(1,XY(1,J))                               MAIN   40
        WRITE(6,27) Z(NR),NP(NR),((XY(K,J),K=1,2),J=KA,KB)    MAIN   41
27      FORMAT('OREGION WITH VALUE',F5.0,'DEFINED BY',F5.0,'BOUNDARY POINTMAIN 42
       1S',/,' COORDINATES:',/,(' ',2F1C.0))                 MAIN   43
        NR=NR+1                                               MAIN   44
        GO TO 24                                              MAIN   45
22      NR=NR-1                                               MAIN   46
```

The main program finishes by calling the appropriate subroutines:

```
        CALL SCALE(XY,KB,NCHARS,NLINES,1)                     MAIN   47
        CALL BDRY2(XY,KB,NP,Z,NR,MAP,NCHARS,NLINES)          MAIN   48
        CALL POINT(XY,KB,NCHARS,NLINES)                       MAIN   49
        CALL MAP2(MAP,NCHARS,NLINES)                          MAIN   50
        STOP                                                  MAIN   51
        END                                                   MAIN   52
```

Note that POINT is called *after* BDRY2. This is because POINT sorts the input data (in this case, XY) before it plots it with the printer. If these data were then sent to BDRY2, the results would be meaningless.

For a main program whose goal is not a computer map with MAP2, but a matrix representing square cells to be used with MAP4 or for further computation, SCALE is called with the last argument set to 0. The column coordinates in XY must be converted to line numbers after BDRY2 is called, and before POINT is called.

Subroutine BDRY2 is based on two assumptions:

1 the row and column coordinates of the boundary points of each region start at the point which is furthest left, and are arranged clockwise from there (thus points on the original map would start at the top and proceed counterclockwise);
2 the map is rectangular, and completely accounted for as regions (though the regions may be in any order).

If either of these assumptions does not hold, then BDRY2 will not produce the desired result.

The coding for BDRY2 commences with comments, specifications, and the initialization of the output matrix MAP:

```
      SUBROUTINE BDRY2(RC,N,NP,Z,NR,MAP,NROWS,NCOLS)          BDY2   1
C FILLS MATRIX WITH VALUES FROM BOUNDARY CODING               BDY2   2
C RC CONTAINS ROW AND COLUMN COORDINATES (ROW IN ROW 1 OF RC, COL IN  BDY2   3
C ROW 2) OF BOUNDARY POINTS, STARTING AT THE FAR LEFT POINT AND BDY2   4
C PROCEEDING CLOCKWISE                                         BDY2   5
C WHERE THE MATRIX IS TRANSPOSE OF SOURCE MAP. COORDINATES ON SOURCE  BDY2   6
C WOULD START AT TOP AND PROCEED COUNTERCLOCKWISE, AND WOULD BE BDY2   7
C TRANSFORMED BY SCALE BEFORE USED HERE                        BDY2   8
C N IS NUMBER OF BOUNDARY POINTS                               BDY2   9
C NP IS ARRAY OF NUMBERS OF POINTS IN EACH REGION              BDY2  10
C Z IS ARRAY OF VALUES OF REGIONS                              BDY2  11
C NR IS NUMBER OF REGIONS                                      BDY2  12
      REAL RC(2,N),NP(NR),Z(NR)                                BDY2  13
C A WILL CONTAIN SORTED COORDINATES AND VALUES (MAX OF 500 BOUNDARY  BDY2  14
C SEGMENTS)                                                    BDY2  15
      REAL A(5,500)                                            BDY2  16
C SAVE IS REQUIRED FOR TEMPORARY STORAGE                       BDY2  17
      REAL SAVE(6,15)                                          BDY2  18
C MAP IS MATRIX NROWS BY NCOLS TO BE FILLED                    BDY2  19
      REAL MAP(NROWS,NCOLS)                                    BDY2  20
C SET ALL MAP ELEMENTS TO 0.                                   BDY2  21
      DO 1 J=1,NCOLS                                           BDY2  22
      DO 1 I=1,NROWS                                           BDY2  23
    1 MAP(I,J)=0.                                              BDY2  24
```

Most of the subroutine is concerned with setting those elements of MAP through which a boundary passes to the value of the region to which the boundary belongs. Since most boundary segments are common to two regions, this operation may be limited to those line segments on the upper side of a region, that is, those in which the column coordinate of the second point is greater than that of the first point (figure 6.4). Since the entire map is accounted for by regions, all boundary segments except those along the bottom margin will be on the upper side of a region.

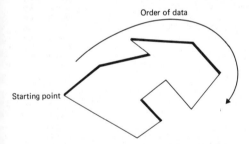

Order of data

Starting point

Figure 6.4 A region showing those line segments considered to be on its upper side as thicker lines. Since the boundary data start at the far left point and are ordered clockwise, the test for this is if the column coordinate at the end of a line segment is greater than the column coordinate of the point at the beginning of that line segment.

All boundary segments are now examined, and those which are on the upper side of a region identified and stored in the array A together with the region's value:

```
C                                                             BDY2  25
C FILL A WITH COORDINATES OF BOUNDARY SEGMENTS ON THE UPPER SIDE  BDY2  26
C OF REGIONS, AND THE REGION'S VALUE                          BDY2  27
      KOUNT=0                                                 BDY2  28
      MB=-1                                                   BDY2  29
      DO 10 L=1,NR                                            BDY2  30
      MA=MB+2                                                 BDY2  31
      MB=MA+NP(L)-2                                           BDY2  32
      DO 10 K=MA,MB                                           BDY2  33
C CHECK IF ON UPPER SIDE                                      BDY2  34
      IF(RC(2,K+1)-RC(2,K).LE.0.) GO TO 10                    BDY2  35
      KOUNT=KOUNT+1                                           BDY2  36
```

```
          A(1,KOUNT)=Z(L)                                          BDY2  37
          A(2,KOUNT)=RC(1,K)                                       BDY2  38
          A(3,KOUNT)=RC(2,K)                                       BDY2  39
          A(4,KOUNT)=RC(1,K+1)                                     BDY2  40
          A(5,KOUNT)=RC(2,K+1)                                     BDY2  41
   10     CONTINUE                                                 BDY2  42
```

The boundary segments are now sorted to ensure that all cells in MAP in which more than one boundary occurs have the value of the one on the lower side. The procedure used for this involves several steps described in the comment cards:

```
C                                                                 BDY2  43
C                                                                 BDY2  44
C SORT A SO THAT LINES ARE ORDERED WITH ROW COORDINATES OF        BDY2  45
C STARTING POINTS IN DESCENDING ORDER, AND FOR LINES WITH THE     BDY2  46
C SAME ROW COORDINATE OF STARTING POINT, THE ROW COORDINATES OF THE  BDY2  47
C END POINT ARE IN DESCENDING ORDER                               BDY2  48
C                                                                 BDY2  49
C SET UP A(2,*) SO THAT IT IS A DECIMAL NUMBER WHOSE FIRST PART IS  BDY2  50
C THE ROW COORDINATE OF THE STARTING POINT, AND WHOSE             BDY2  51
C DECIMAL PART IS THE ROW COORDINATE OF THE END POINT             BDY2  52
C ASSUME THERE ARE NO MORE THAN 2 SIGNIFICANT DIGITS AFTER THE DECIMAL  BDY2  53
C IN THE ROW COORDINATE OF THE STARTING POINT                    BDY2  54
          DO 13 K=1,KOUNT                                         BDY2  55
          A(2,K)=IFIX(A(2,K)*100.)+A(4,K)/(NROWS+1)               BDY2  56
C REVERSE ORDER SO THAT SUBROUTINE SORT CAN BE USED               BDY2  57
   13     A(2,K)=-A(2,K)                                          BDY2  58
          CALL SORT(A,KOUNT,5,2)                                  BDY2  59
C RESET A(2,*)                                                    BDY2  60
          DO 14 K=1,KOUNT                                         BDY2  61
          A(2,K)=-IFIX(A(2,K))                                    BDY2  62
   14     A(2,K)=A(2,K)/100.                                      BDY2  63
C                                                                 BDY2  64
C LINES WITH IDENTICAL ROW COORDINATES OF START AND END POINTS    BDY2  65
C ARE NOW IDENTIFIED AND SORTED SO THAT UPWARD SLOPING LINES      BDY2  66
C ARE ORDERED FROM RIGHT TO LEFT, AND DOWNWARD SLOPING LINES      BDY2  67
C FROM LEFT TO RIGHT                                              BDY2  68
C ASSUME NO MORE THAN 15 LINES WITH THE SAME ROW COORDINATES      BDY2  69
          K=1                                                     BDY2  70
  131     K=K+1                                                   BDY2  71
          IF(K.EQ.KOUNT) GO TO 140                                BDY2  72
          DO 132 L=K,KOUNT                                        BDY2  73
          IF(A(2,K-1).NE.A(2,K)) GO TO 133                        BDY2  74
          IF(A(4,K-1).NE.A(4,K)) GO TO 133                        BDY2  75
  132     CONTINUE                                                BDY2  76
  133     IF(K.EQ.L) GO TO 131                                    BDY2  77
C IF LINES ARE HORIZONTAL, TAKE NO ACTION                         BDY2  78
          IF(A(3,K).EQ.A(5,K)) GO TO 136                          BDY2  79
C                                                                 BDY2  80
C PUT THIS SECTION OF A INTO THE FIRST FIVE ROWS OF SAVE          BDY2  81
C IN THE SIXTH ROW OF SAVE PUT THE SUM OF THE COLUMN COORDINATES, SET  BDY2  82
C NEGATIVE IF LINES ARE DOWNWARD SLOPING, OTHERWISE POSITIVE      BDY2  83
          K1=K-1                                                  BDY2  84
          IJ=0                                                    BDY2  85
          SLOPE=1.                                                BDY2  86
          IF(A(4,K).LT.A(2,K)) SLOPE=-1.                          BDY2  87
          DO 134 M=K1,L                                           BDY2  88
          IJ=IJ+1                                                 BDY2  89
          DO 135 J=1,5                                            BDY2  90
  135     SAVE(J,IJ)=A(J,M)                                       BDY2  91
  134     SAVE(6,IJ)=SLOPE*(A(3,M)+A(5,M))                        BDY2  92
C NOW SORT                                                        BDY2  93
          CALL SORT(SAVE,L-K1+1,6,6)                              BDY2  94
C PUT SORTED VALUES BACK INTO A                                   BDY2  95
          IJ=0                                                    BDY2  96
          DO 137 M=K1,L                                           BDY2  97
          IJ=IJ+1                                                 BDY2  98
          DO 137 J=1,5                                            BDY2  99
  137     A(J,M)=SAVE(J,IJ)                                       BDY2 100
  136     K=L-1                                                   BDY2 101
          GO TO 131                                               BDY2 102
  140     CONTINUE                                                BDY2 103
```

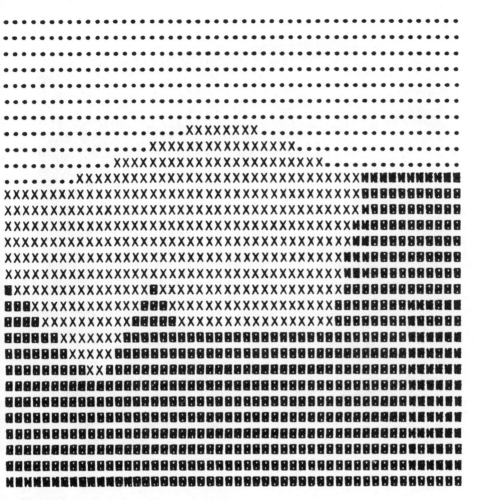

Figure 6.5 Some putput from BDRY2 and MAP2.

Now each matrix element in which a boundary segment is set to the value of the region below that boundary.

```
C                                                                    BDY2 104
C FILL ELEMENTS OF MATRIX IN WHICH A BOUNDARY FALLS WITH VALUE OF     BDY2 105
C REGION BELOW IT                                                     BDY2 106
      DO 20C L=1,KOUNT                                                BDY2 107
C ICOL AND JCOL ARE NUMBERS OF FIRST AND LAST COLUMNS IN WHICH        BDY2 108
C THIS SEGMENT OCCURS                                                 BDY2 109
      ICOL=A(3,L)                                                     BDY2 110
      JCOL=A(5,L)                                                     BDY2 111
C DROW1 IS ROW DISTANCE PER UNIT COLUMN                               BDY2 112
      DROW1=(A(4,L)-A(2,L))/(JCOL-ICOL)                               BDY2 113
C INITIALIZE ROW                                                      BDY2 114
      ROW=A(2,L)-DROW1                                                BDY2 115
      DO 200 J=ICOL,JCOL                                              BDY2 116
      ROW=ROW+DROW1                                                   BDY2 117
C SET LOCATION IN MAP (UNLESS ALREADY ON A BOUNDARY)                  BDY2 118
200   IF(MAP(ROW,J).EQ.0.)MAP(ROW,J)=A(1,L)                          BDY2 119
```

The map matrix is now filled with the region values by examining the matrix cell immediately above each cell.

Plate 4   A drum plotter. (Photograph provided by California Computer Products, Inc.)

```
C                                                           BDY2 120
C FILL MAP WITH VALUES                                      BDY2 121
       DO 400 J=1,NCOLS                                     BDY2 122
       DO 400 I=2,NROWS                                     BDY2 123
       IF(MAP(I,J).EQ.0.) MAP(I,J)=MAP(I-1,J)               BDY2 124
400    CONTINUE                                             BDY2 125
       RETURN                                               BDY2 126
       END                                                  BDY2 127
```

6.3   Line Plotting Machines

As useful as they are, maps produced by a standard computer line printer provide little competition for the professional cartographer. It is possible, however, to produce good quality graphic output with a computer by using special line plotting devices. The most common of these machines fall into five categories:

1   **Cathode Ray Tubes (CRT)**   The cathode ray tube operates on the same principle as television, and looks much like one. Graphic output is displayed on the CRT screen by the computer, and either renewed many times a second (a refresh scope), or held constant on the screen (a storage scope). These machines are invaluable for interactive work with the main computer because of their speed (a new picture can be put

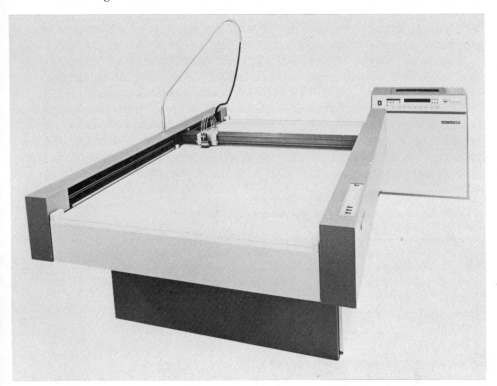

Plate 5   A flatbed plotter. (Photograph provided by California Computer Products, Inc.)

up in less than a second) and the ease with which the user can transmit commands (either through a typewriter console or with a device identifying a place on the display). Their principal disadvantages for spatial problems are their low resolution and the difficulty in getting the screen output transferred to paper—the usual method is to take a photograph. (The distinction is made when working with CRT between softcopy and hardcopy. The former is what you see displayed on the screen; the latter is a copy of this on film or paper.) Most medium to large computer centres will have one or more CRT's, although not always with plotting capability.

2   **Drum Plotters**   The drum plotter is a relatively low cost, high resolution (typically 1/200 in) machine which produces ball point or ink plots from one to three feet wide, and up to several feet long. Plate 4 is a photograph of such a machine. It operates by rotating a drum and moving a penholder across the plotting area under program control, with the plotting pen either up or down. Since this movement is done with stepping motors, the pen can only move in one of eight directions, and lines in certain directions show slight irregularities under close examination. Drum plotters are the most commonly available graphic output providing good quality output.

3   **Microfilm Recorders**   An electron beam plots lines on microfilm in this machine. The output is either enlarged and printed, or viewed in a microfilm reader. This type of machine is particularly useful when many plots must be produced and saved.

4   **Flatbed Plotters**   These machines are similar to drum plotters except the plotting surface is a table (plate 5). They tend to provide better resolution than drum plotters, and much more flexibility in linewidths, colours, and the possibility of editing during

the plotting operation. These are probably the best machines for general purpose automated map production, but are not available even in most large computer centres.

5   **Electrostatic Plotters**   The principle of electrostatic plotting (as in a Xerox machine) is used in some output devices which do both printing and plotting at extremely high speeds, but with relatively low resolution and contrast.

There are several other kinds of machines which produce graphic output, but thus far they are very few in number. The reader interested in these specialized machines or in additional information on more conventional equipment is advised to read the chapters by A R Boyle in Tomlinson (1972). A very general comparison of the five types of machine described above (plus the standard line printer) is presented in table 6.1.

Table 6.1   Comparison of Graphic Output Machines

	Availability	Resolution	Speed	Cost (per plot)
Cathode ray tubes	common	low	high	low to medium
Drum plotters	common	high	low	medium to high
Microfilm recorders	not common	high	high	medium
Flatbed plotters	rare	high	low	high
Electrostatic plotters	rare	low	high	low
Line printer	most common	low	high	low

6.4   Plotters and Spatial Data

Any computer mapping that can be done with a line printer can be done more accurately and usually more attractively with a line plotter. The additional cost, however, is substantial. Programs are more difficult to write and take longer to run; special charges are assessed for plotter use, and turn-around time is typically hours rather than minutes. These costs are sufficiently high that it is usually not worth using a plotter to produce maps which are to be used only for a short period of time, or are to be studied by individuals familiar with line printer output. If the maps are to be preserved for future use, presented to individuals unfamiliar with computer output, or published, the additional cost and inconvenience of plotter output is usually warranted.

One example is presented in this book which illustrates a typical plotter program. Example 14 is a main program and subroutine which plots map outlines and graduated circles at specified locations. There are two other kinds of operations on spatial data which are done with a plotter—drawing isarithmic maps and perspective views. No FORTRAN coding is presented for either of these because it tends to be long, and there are very good programs already available for these tasks at most computer centres. Some general comments are presented below, however, which describes the computation procedure for each operation.

The usual method for producing isarithmic maps on a plotter starts with a matrix whose elements represent values at intersections on a grid. If the original data are irregularly spaced, this matrix must first be computed by one of the two methods described in chapter 5, surface fitting or numerical approximation. These values are then examined in groups of four and the positions of end points of isarithms determined by linear interpolation. Thus, in figure 6.6, the distance a is given by

$$a = \frac{40 - 33}{45 - 33} \times \text{side}$$

One of the principal limitations of plotter contours produced in this way is that they tend to have a ragged appearance because they are made up of straight line segments whose length is dependent on the grid size. Modifying these to curves is difficult, but is a feature available in some programs.

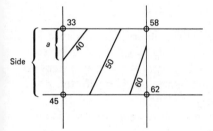

Figure 6.6   The position of an isarithm within a grid cell.

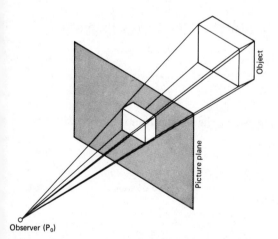

Figure 6.7   Perspective view of an object.

In order to produce a perspective of a surface or object on a line plotter, it is necessary to apply a series of mathematical transformations to its three-dimensional coordinates. Instead of a surface, let us consider an object which has been digitized by the coordinates of the end points of line segments (figure 6.7). We assume that a viewer of this object is situated at point P_0, and is looking at the object at point P_1 (figure 6.8).

The first step is a translation of all the coordinates so that the line of sight from P_0 to P_1 is made to correspond with one of the coordinate axes, and point P_1 becomes

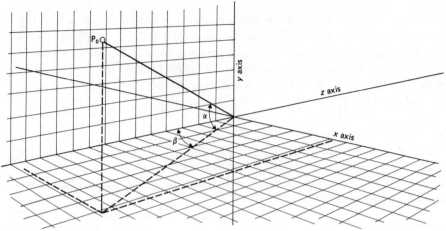

Figure 6.8 Rotation angles.

the origin.[1] This translation moves both the object and the observer the distance required to shift P_1 to the origin, and is accomplished simply by subtracting P_1 from all points defining the object and from the observer's location. The next two steps are rotations of all points in the object so that the picture plane coincides with the x–y plane and the line of sight with the z axis (figure 6.8).

The elements of the rotation matrix $\mathbf{R}_{xy}$ in the general case where rotation is the angle α in the two-dimensional plane x–y are

$$r_{ii}=1 \qquad \text{except } r_{xx}=r_{yy}=\cos\alpha$$
$$r_{ij}=0 \qquad \text{except } r_{xy}=r_{yx}=-\sin\alpha$$

If α is defined as the angle of rotation required about the x axis, the first rotation matrix is

$$\mathbf{R}_{yz}=\begin{bmatrix} 1 & 0 & 0 \\ 0 & \cos\alpha & \sin\alpha \\ 0 & -\sin\alpha & \cos\alpha \end{bmatrix}$$

and if β is the angle of rotation about the y axis, the second rotation matrix is

$$\mathbf{R}_{xy}=\begin{bmatrix} \cos\beta & 0 & -\sin\beta \\ 0 & 1 & 0 \\ \sin\beta & 0 & \cos\beta \end{bmatrix}$$

The two rotations may be combined to give

$$\begin{bmatrix} x_i \\ y_i \\ z_i \end{bmatrix}=\begin{bmatrix} \cos\beta & 0 & -\sin\beta \\ \sin\alpha\sin\beta & \cos\alpha & \sin\alpha\cos\beta \\ \cos\alpha\sin\beta & -\sin\alpha & \cos\alpha\cos\beta \end{bmatrix}\begin{bmatrix} x_i' \\ y_i' \\ z_i' \end{bmatrix}$$

where x_i', y_i' and z_i' are the coordinates after translation, and x_i, y_i and z_i are the coordinates after translation and rotation.

After the above operation, the coordinate system of the object has been transformed so that the observer is at $(0, 0, d)$, where d is the distance of the observer to the picture plane. It is now necessary to determine the x and y coordinates of each

---

[1] The language and notation which follow are those of matrix algebra. Those readers with no background in this subject may wish only to examine the figures.

projected point p $(x, y, 0)$ in the picture plane (figure 6.9). These are given by

$$x_i'' = x_i\{d/(d-z_i)\}$$
$$y_i'' = y_i\{d/(d-z_i)\}$$

These mathematical transformations are not difficult to code in a FORTRAN program because they are simply a series of assignment statements. Significant problems do arise, however, when one attempts to remove lines which should be hidden by that part of the object or surface which is closer to the observer. This is a difficult computing problem which commonly requires much more time than the actual perspective calculations.

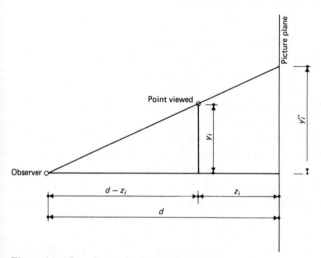

Figure 6.9   Coordinates in the picture plane.

6.5   Programming the Plotter

The specific procedures by which a programmer includes line plotting in a computer program depend on the machine and the conventions in effect at the computer installation. In general four kinds of operations are required:

1   A compiler must be specified which includes plotter capability, that is, makes available the subroutines required to control the plotter, and either attaches the plotter to the main computer, or attaches a tape unit which will then be used to drive the plotter. (The latter procedure is termed 'off-line' plotting.)
2   The plotter must be initialized, a collection of tasks including allocating a special area of core storage to contain plotter instructions (termed 'buffers'), plotting the user's name and other identification (or writing this on the plot tape), and specifying maximum dimensions of the plot (in the case of a drum plotter, the maximum length of plot). This initializing can usually be done with a call to a library subroutine at the outset of the program.
3   Plotting instructions within the program are calls to library subroutines which instruct the pen to move in a certain way, connect points, plot character information, or carry out more complicated operations such as drawing and labelling the axis of a graph. Some installations offer dozens of such subroutines, and often

many of these are unique to that particular system (having been written by programmers at that centre). However, there are always three or four basic routines: one to move the pen or light beam to a new coordinate location, either drawing a line or not (usually called PLOT), one to connect a series of points with a line (usually called LINE), and one to plot numerical or character information at a point (sometimes two different routines are required for this).

4 The plot must be terminated. This is usually done with a call to a subroutine which disconnects the plotter (or writes a mark on the plotter tape), and signals the plotter operator. The following example illustrates the last three of these steps.

6.6   Example 14. CIRCLE: Graduated Circles

One type of map for which the line plotter is ideally suited is the graduated symbol map, in which the size of each symbol is proportional to the value at a location or within a region. These maps are most useful when the quantities symbolized are of more interest than details of location, for example, the population of large cities, or the tonnage shipped from large ports.

The circle is by far the most frequently used symbol for such maps, and has been for over a century. The usual procedure is to make the area of the circles proportional to the values being mapped, with appropriate scaling factors chosen to keep circle sizes within a reasonable range on the map. The drawback to this approach is that the ordinary observer tends to significantly underestimate the area of larger circles in relation to smaller ones. The visual significance of the smaller circles is thus increased.

To compensate for this visual effect, Robinson and Sale (1969) suggest the following transformation of the data:

$$x' = x^{0.57}$$

This is a relatively complicated computation when done manually, requiring the use of logarithmic tables. In a computer, it is a simple operation.

The difficult part of this problem is that plotters draw straight lines, not curves. This means that the computer program must determine the coordinates of points along the circumference of each circle which, when joined together by straight line segments, will give the impression of a smooth curve to the ordinary observer. The equation of a circle with centre at 0, 0 is

$$x^2 + y^2 = r^2$$

where r is the radius. This can be rewritten as

$$x = (r^2 - y^2)^{0.5}$$

and used in difference form to compute coordinates for one quarter of the circumference:

```
      NN=RADIUS/RES+1.
      RADSQ=RADIUS**2
      DO 125 K=1,NN
      Y(K)=(K-1)*RES
125   X(K)=SQRT(RADSQ-Y(K)**2))
```

where RADIUS is the radius of the circle and RES is the length of the shortest straight line segment which can be tolerated in the plot. Coordinates for the other three quarters of the circumference are the same as those computed by the above loop, except that the signs are changed: Y negative and X positive for the lower right quarter, both X and Y negative for the lower left, and X negative and Y positive for the upper left quarter.

Most plotting systems are operated by two library subroutines which initialize and terminate the plot, plus a number of subroutines which control the plotting (linking points, drawing symbols, letters, axes of graphs, etc.). In this example, we will assume only the initializing and terminating subroutines (called BEGIN and FINISH), and a subroutine which connects points with a straight line (PLINE).

A main program is presented below which calls BEGIN and FINISH, reads and plots map outline information, reads data values and circle centres, and calls CIRCLE. These are relatively simple operations, but the FORTRAN coding is lengthy because, as in example 13, it is necessary to allow for strings of numbers of various lengths. The data are assumed to be in three parts, each terminated by a fence card:

1   the coordinates of outline points;

2   for each part of the outline, a card giving the total number of outline points, and one or more cards with the numbers of the outline points;

3   the coordinates of the circle centres and the value of each.

Figure 6.10 is an example of some data arranged in this way.

All coordinates are of the *x* and *y* form (rather than row and column), and are assumed to be scaled to fit the available plotter area. If they are not, most plotter software includes a scaling subroutine.

The FORTRAN statements for reading the various parts of the data are presented below, together with the required specifications and the initial call to BEGIN.

```
   0.0        0.0 ⎤
   6.5        0.0 ⎥ coordinates of outline points
   6.5        8.0 ⎥
   0.0        8.0 ⎦
99999.0────────────── fence card
     5
1    2    3    4    1 ⎫ points comprising outline
                     ⎭
99999────────────────── fence card
     1        2    14.7 ⎤
     2        3   219.4 ⎥
     3        4   175.2 ⎥ coordinates of centres and values of circles
     4        5    74.1 ⎥
     5        6   100.0 ⎦
99999────────────────── fence card
```

Figure 6.10   A data set arranged in the format required by the main program for CIRCLE. The outline consists of a rectangle 6.5 by 8.0 inches.

```
C MAIN PROGRAM FOR CIRCLE                                        MAIN   1
      DIMENSION X(100),Y(100),Z(100),VERT(2,100)                 MAIN   2
C INITIALIZE PLOTTER                                             MAIN   3
      CALL BEGIN                                                 MAIN   4
C READ COORDINATES OF OUTLINE POINTS UNTIL A FENCE CARD ENCOUNTERED  MAIN   5
      NVERT=1                                                    MAIN   6
12    READ(5,10)(VERT(J,NVERT),J=1,2)                            MAIN   7
10    FORMAT(2F10.0)                                             MAIN   8
      IF(VERT(1,NVERT).EQ.99999.)GO TO 11                        MAIN   9
      NVERT=NVERT+1                                              MAIN  10
      GO TO 12                                                   MAIN  11
11    NVERT=NVERT-1                                              MAIN  12
      WRITE(6,13)                                                MAIN  13
13    FORMAT('1COORDINATES OF OUTLINE POINTS READ:')             MAIN  14
      WRITE(6,20)(K,(VERT(J,K),J=1,2),K=1,NVERT)                 MAIN  15
20    FORMAT('0',I5,2F10.2)                                      MAIN  16
C READ VERTEX NUMBERS MAKING UP LINE SEGMENTS UNTIL A FENCE CARD MAIN  17
C ENCOUNTERED                                                    MAIN  18
24    READ(5,21)NPTS                                             MAIN  19
```

```
21      FORMAT(I10)                                            MAIN  20
        IF(NPTS.EQ.99999)GO TO 22                              MAIN  21
        READ(5,23) (Z(I),I=1,NPTS)                             MAIN  22
23      FORMAT(14F5.0)                                         MAIN  23
        DO 26 J=1,NPTS                                         MAIN  24
        Y(J)=VERT(2,Z(J))                                      MAIN  25
26      X(J)=VERT(1,Z(J))                                      MAIN  26
C PLOT OUTLINE                                                 MAIN  27
        CALL PLINE(X,Y,NPTS)                                   MAIN  28
        GO TO 24                                               MAIN  29
C READ COORDINATES OF CIRCLE CENTERS AND VALUES UNTIL A FENCE  MAIN  30
C CARD ENCOUNTERED                                             MAIN  31
22      NCIRC=1                                                MAIN  32
40      READ(5,41) X(NCIRC),Y(NCIRC),Z(NCIRC)                  MAIN  33
41      FORMAT(3F10.0)                                         MAIN  34
        IF(X(NCIRC).EQ.99999.) GO TO 42                        MAIN  35
        NCIRC=NCIRC+1                                          MAIN  36
        GO TO 40                                               MAIN  37
42      NCIRC=NCIRC-1                                          MAIN  38
        WRITE(6,43)                                            MAIN  39
43      FORMAT('0COORDINATES AND SIZES OF CIRCLES:')           MAIN  40
        WRITE(6,44) (X(I),Y(I),Z(I),I=1,NCIRC)                 MAIN  41
44      FORMAT(' ',3F10.0)                                     MAIN  42
```

It is assumed that the data which are to be plotted as graduated circles have not been scaled. This is done in CIRCLE, using arguments set in the CALL which specify the minimum and maximum radius into which the range of circle sizes is to be fitted, and the resolution desired in the plot, that is, the increment in x or y used to compute the coordinates of points defining the circumference of the circles. These are the last three arguments in the CALL statement below. When control is returned to the main program, FINISH is called to terminate the plotting.

```
        CALL CIRCLE(X,Y,Z,NCIRC,1.5,0.3,0.1)                   MAIN  43
C TERMINATE PLOT                                               MAIN  44
        CALL FINISH                                            MAIN  45
        STOP                                                   MAIN  46
        END                                                    MAIN  47
```

Subroutine CIRCLE starts with comments and specifications, then scales the data values to radii which will be used in plotting the graduated circles:

```
        SUBROUTINE CIRCLE(X,Y,Z,NCIRC,RMAX,RMIN,RES)           CIRC   1
C PLOTS GRADUATED CIRCLES WITH CENTERS AT X,Y WITH RADII PROPORTIONAL  CIRC   2
C TO Z VALUES ACCORDING TO     R=Z**0.57                       CIRC   3
C CIRCLES ARE SCALED TO FIT BETWEEN SPECIFIED MINIMUM AND MAXIMUM  CIRC   4
C RADII (RMIN AND RMAX)                                        CIRC   5
        DIMENSION X(NCIRC),Y(NCIRC),Z(NCIRC)                   CIRC   6
C A AND B ARE ARRAYS FOR CIRCLE PLOTTING;ASSUME MAXIMUM OF 400 POINTS  CIRC   7
C PER CIRCLE, AND REUSE A & B FOR EACH QUARTER OF CIRCUMFERENCE CIRC   8
        DIMENSION A(100),B(100)                                CIRC   9
C RES IS MAXIMUM DISTANCE IN X OR Y BETWEEN POINTS ON CIRCUMFERENCE  CIRC  10
C CONVERT Z'S AND FIND MAX AND MIN                             CIRC  11
        ZMAX=Z(1)**0.57                                        CIRC  12
        ZMIN=Z(1)**0.57                                        CIRC  13
        DO 100 I=2,NCIRC                                       CIRC  14
        Z(I)=Z(I)**0.57                                        CIRC  15
        IF(Z(I).GT.ZMAX) ZMAX=Z(I)                             CIRC  16
        IF(Z(I).LT.ZMIN) ZMIN=Z(I)                             CIRC  17
100     CONTINUE                                               CIRC  18
C SCALE Z'S                                                    CIRC  19
        FACTOR=(RMAX-RMIN)/(ZMAX-ZMIN)                         CIRC  20
        DO 200 I=1,NCIRC                                       CIRC  21
200     Z(I)=(Z(I)-ZMIN)*FACTOR+RMIN                           CIRC  22
```

The circles are plotted in quarters so that the same arrays (A and B) can be used four times. This saves memory and computing time, since arithmetic operations are much faster than functions such as SQRT. The first step is to determine the number of increments of length RES required in either X or Y to plot one eighth of the

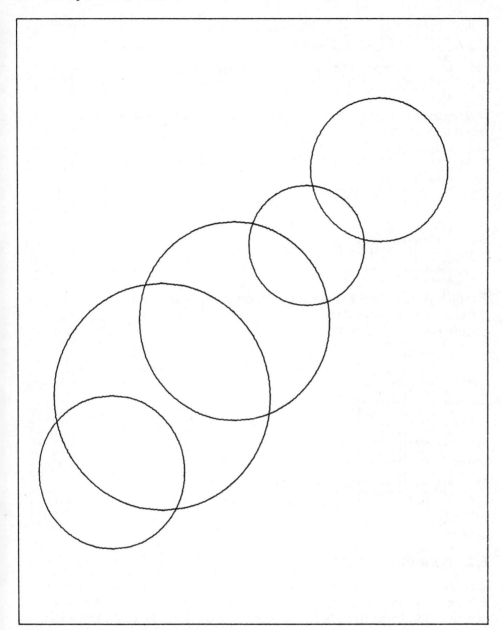

Figure 6.11   Plotter output from example 14.

circumference using the relationship that one eighth of the circumference when projected onto the x or y axis subtends two thirds of the radius. Twice this value is the total number of elements required for plotting one quarter of the circumference of the Kth circle. This is checked to ensure it is less than the specified maximum:

```
      DO 500 K=1,NCIRC                                               CIRC  23
C CALCULATE NUMBER OF INCREMENTS TO PLOT ONE-EIGHTH OF CIRCUMFERENCE CIRC  24
      N8=(Z(K)/RES)*2./3.+1                                          CIRC  25
      N4=N8*2.-1.                                                    CIRC  26
```

```
C CHECK CIRCLE SIZE                                                CIRC  27
      IF(N4.LT.100) GO TO 250                                      CIRC  28
C WRITE ERROR MESSAGE AND RETURN                                   CIRC  29
      WRITE(6,251)K                                                CIRC  30
251   FORMAT('0CIRCLE',I3,' IS TOO LARGE. RESET RES TO LARGER VALUE, DECCIRC  31
     *REASE CIRCLE SIZE, OR REDUCE RMAX.')                         CIRC  32
      CALL FINISH                                                  CIRC  33
      STOP                                                         CIRC  34
```

The coordinates of the circle, but with an origin at 0, 0, are then computed and stored in A and B:

```
250   R2=Z(K)*Z(K)                                                 CIRC  35
C USE INCREMENTS OF RES UNITS IN Y FIRST                           CIRC  36
      DO 290 I=1,N8                                                CIRC  37
      B(I)=(I-1)*RES                                               CIRC  38
290   A(I)=SQRT(R2-B(I)*B(I))                                      CIRC  39
C NOW USE INCREMENTS OF RES UNITS IN X                             CIRC  40
      N8=N8+1                                                      CIRC  41
      DO 300 I=N8,N4                                               CIRC  42
      A(I)=A(I-1)-RES                                              CIRC  43
300   B(I)=SQRT(R2-A(I)*A(I))                                      CIRC  44
C ADD ONE POINT TO CLOSE CIRCLE AT TOP AND BOTTOM                  CIRC  45
      N4=N4+1                                                      CIRC  46
      B(N4)=Z(K)                                                   CIRC  47
      A(N4)=0.                                                     CIRC  48
```

For each quarter of the Kth circle, the vectors A and B are changed as required, and the plotting subroutine called:

```
C PLOT UPPER RIGHT QUARTER                                         CIRC  49
      DO 301 I=1,N4                                                CIRC  50
      A(I)=A(I)+X(K)                                               CIRC  51
301   B(I)=B(I)+Y(K)                                               CIRC  52
      CALL PLINE(A,B,N4)                                           CIRC  53
C PLOT LOWER RIGHT QUARTER                                         CIRC  54
      DO 302 I=1,N4                                                CIRC  55
302   B(I)=Y(K)-(B(I)-Y(K))                                        CIRC  56
      CALL PLINE(A,B,N4)                                           CIRC  57
C PLOT LOWER LEFT QUARTER                                          CIRC  58
      DO 303 I=1,N4                                                CIRC  59
303   A(I)=X(K)-(A(I)-X(K))                                        CIRC  60
      CALL PLINE(A,B,N4)                                           CIRC  61
C PLOT UPPER LEFT QUARTER                                          CIRC  62
      DO 304 I=1,N4                                                CIRC  63
304   B(I)=Y(K)+Y(K)-B(I)                                          CIRC  64
      CALL PLINE(A,B,N4)                                           CIRC  65
500   CONTINUE                                                     CIRC  66
      RETURN                                                       CIRC  67
      END                                                          CIRC  68
```

6.7   Networks

Networks are not usually represented in a computer by their actual geometry, that is, locations of nodes and lengths of links, rather they are coded as a matrix in which each row and column represents some property of the link between the two nodes such as the presence of a direct connection, travel time, or volume of traffic. This matrix is often termed a graph of the network, since the field of mathematics concerned with analysing such patterns is graph theory.

Figure 6.12 is an example of a simple network, and table 6.2 is the corresponding connectivity matrix, that is, a matrix in which the presence of a link or *edge* between two nodes or *vertices* is symbolized by a 1, and the absence of a direct link by a 0. This matrix is symmetrical, that is, an element at a given row and column position in the upper right half is the same as the element in the corresponding column and row position in the lower left half. For example, there is a 1 at row B column E, and also

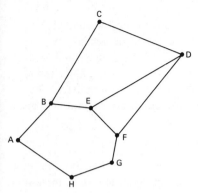

Figure 6.12   A hypothetical network.

at row E, column B. Elements along the diagonal are set to 0, there being no link or edge between a node and itself.

In graphs which are directed, that is, in which some links are one-way, the matrix is asymmetrical. For example, if the link between D and E functioned only from D to E, and not the reverse, then there would be a 1 at row D, column E, but a 0 at row E, column D.

The convention is that movement is from nodes on the left side of the matrix to nodes listed along the top.

Two types of network problems are usefully treated in computers:

1   routing, that is, determining optimal or least cost paths between sets of vertices; and
2   flow, that is, determining the flow capacity of part or all of a network, and ways in which this capacity may most easily be increased or decreased.

Problems of both types are extraordinarily difficult or not feasible to solve manually for any but the smallest networks, but are often easily resolved by computer calculations.

Table 6.2   Connectivity Matrix (for figure 6.12)

	A	B	C	D	E	F	G	H
A	0	1	0	0	0	0	0	1
B	1	0	1	0	1	0	0	0
C	0	1	0	1	0	0	0	0
D	0	0	1	0	1	1	0	0
E	0	1	0	1	0	1	0	0
F	0	0	0	1	1	0	1	0
G	0	0	0	0	0	1	0	1
H	1	0	0	0	0	0	1	0

A comprehensive description of the algorithms which have been formulated for these problems is beyond the scope of this book. (The interested reader may wish to consult the references listed at the end of the book.) One example is presented below, however, which illustrates the compactness of many network processing algorithms, and the advantage of the matrix form for representing networks in computers. This example calculates a matrix of the shortest distances between all nodes on a network.

6.8   Example 15. SHORT: Minimal Distance in a Network

SHORT operates by comparing the lengths of alternative paths between all nodes in a network. It proceeds systematically, enumerating and evaluating all two-link, three-link, four-link, etc. paths until all possibilities have been tested.

The calculations are carried out on the structure matrix, that is, a graph of the network which contains distances (times, flows, etc.) for pairs of points directly connected, zero on the diagonal, and infinity for pairs of points not directly connected. Table 6.3 is the structure matrix for the network in figure 6.12. The lengths of all possible two-link paths between nodes are determined by adding matching elements of rows and columns in the structure matrix. For example, for nodes C and E in table 6.3:

$$
\begin{bmatrix} \infty \\ 4.4 \\ 0 \\ 4.1 \\ \infty \\ \infty \\ \infty \\ \infty \end{bmatrix}
+
\begin{bmatrix} \infty \\ 1.9 \\ \infty \\ 4.9 \\ 1.8 \\ 0 \\ 1.3 \\ \infty \end{bmatrix}
=
\begin{bmatrix} \infty \\ 6.3 \\ \infty \\ 9.0 \\ \infty \\ \infty \\ \infty \\ \infty \end{bmatrix}
$$

The minimum value in the resulting vector (6.3) is the shortest distance between C and E by a two-step path (via B). This calculation is repeated for all pairs of nodes, then

Table 6.3   Structure Matrix (for figure 6.12)

	A	B	C	D	E	F	G	H
A	0	2.4	∞	∞	∞	∞	∞	3.0
B	2.4	0	4.4	∞	1.9	∞	∞	∞
C	∞	4.4	0	4.1	∞	∞	∞	∞
D	∞	∞	4.1	0	4.9	4.8	∞	∞
E	∞	1.9	∞	4.9	0	1.8	∞	∞
F	∞	∞	∞	4.8	1.8	0	1.3	∞
G	∞	∞	∞	∞	∞	1.3	0	2.0
H	3.0	∞	∞	∞	∞	∞	2.0	0

the resulting matrix is compared element by element with the structure matrix by
$$V_{ij} = \min(U_{ij}, S_{ij})$$
where **U** is the matrix of shortest distances by two-step paths, **S** is the structure matrix, and **V** is a matrix of shortest distances between nodes by either one or two-step paths.

The shortest distance matrix for three-link paths is calculated from the matrix for two-link paths and the original structure matrix. It is then compared with **V**, the matrix of shortest distances by either one- or two-link paths, and a new matrix generated which contains shortest distances by one-, two-, or three-link paths. These calculations are repeated as many times as there are nodes in the network.

This procedure can be described in general form using three matrices: the structure matrix **S**, the shortest distance matrix **U**, and a matrix **W** which contains the shortest distances computed in the previous step in the computation. The elements of **U** are determined by
$$U_{ij} = \min(S_{i1} + W_{1j}, S_{i2} + W_{2j}, S_{i3} + W_{3j}, \ldots, S_{in} + W_{nj})$$
The operation is repeated either until there is no change in the shortest distance matrix **U**, or until all possible combinations of links have been considered.

If coded in the simplest and most direct fashion, a computer program for this procedure requires N^3 operations at each step (where N is the number of nodes). Ackoff and Sasieni (1968) suggest an alternative version which significantly reduces computation time required, and the proportion of time saved over the direct version is proportional to the size of the problem, that is, larger problems have proportionately larger savings. They suggest that the operations described above be conceived of as a special kind of matrix multiplication. This means that determining a matrix of shortest distances using two-step routes is in a sense equivalent to squaring the structure matrix. If this squared structure matrix is then multiplied by itself, the result is the fourth power of the structure matrix, and it contains all the shortest distances between all pairs of nodes in four steps or fewer. If this matrix is again multiplied by itself, the result is the eighth power of the structure matrix ($\mathbf{S}^4 \times \mathbf{S}^4 = \mathbf{S}^8$). This multiplied by itself would be the sixteenth power, and it would contain the shortest distances between pairs of nodes in 16 steps or less. The procedure suggested by Ackoff and Sasieni is to continue multiplying the matrix by itself until there is no further change in it, or until S^N is reached.

SHORT has been written using this faster algorithm. Its arguments are a three-dimensional array S, whose first level, $S(*, *, 1)$ contains the structure matrix. On return from SHORT, the solution is contained in $S(*, *, KK)$ where KK will have the value 1 or 2. Within the subroutine, solutions are calculated and stored alternately in $S(*, *, 1)$ and $S(*, *, 2)$. Because of the algorithm used, all possible paths between nodes can be tested only when the number of nodes is a power of two. If this is not the case, it is advisable to pad the input array $S(*, *, 1)$ with dummy nodes not directly connected to any others until the number of nodes is a power of 2.

The FORTRAN coding starts with specification statements, comments, and assignment statements which initialize several variables:

```
      SUBROUTINE SHORT(S,N,KK)                                    SHRT    1
C MINIMAL DISTANCE ROUTINE                                        SHRT    2
C S(*,*,1) IS INPUT DATA (GRAPH OF N NODE NETWORK)                SHRT    3
C A SOLUTION GUARANTEED ONLY IF N IS A POWER OF 2.  IF N NOT A    SHRT    4
C POWER OF 2, SOLUTION IS CHECKED.  IF INCORRECT, KK IS SET TO 0 AND  SHRT    5
C A MESSAGE IS PRINTED.                                           SHRT    6
C ELEMENTS ARE DISTANCES IF NODES DIRECTLY CONNECTED              SHRT    7
```

```
C  0 IF ON DIAGONAL, 99999 OTHERWISE (INFINITY)                           SHRT   8
C  SOLUTION IS RETURNED IN S(*,*,KK)                                      SHRT   9
       REAL S(N,N,2)                                                      SHRT  10
       BIG=99999.                                                         SHRT  11
       K=1                                                                SHRT  12
       KK=2                                                               SHRT  13
       KOUNT=1                                                            SHRT  14
```

In the block of statements which calculates the solution matrix, KOUNT is the current power of 2, K the level of S which contains the previous solution, and KK the level of S which contains the current solution being computed.

```
C  SQUARE MATRIX                                                          SHRT  15
2       KOUNT=KOUNT+KOUNT                                                 SHRT  16
        IF(KOUNT.EQ.N)RETURN                                             SHRT  17
        IF(KOUNT.GT.N)GO TO 10                                           SHRT  18
        DO 3 J=1,N                                                        SHRT  19
        DO 3 I=1,N                                                        SHRT  20
        TMIN=BIG                                                          SHRT  21
        DO 4 L=1,N                                                        SHRT  22
        SS=S(I,L,K)+S(L,J,K)                                              SHRT  23
4       TMIN=AMIN1(SS,TMIN)                                               SHRT  24
3       S(I,J,KK)=TMIN                                                    SHRT  25
        SAVE=K                                                            SHRT  26
        K=KK                                                              SHRT  27
        KK=SAVE                                                           SHRT  28
        GO TO 2                                                           SHRT  29
```

If N is a power of 2, then the IF statement on card 17 will return control to the calling program. If the second IF is true, however, it means that N is not a power of 2, and execution passes to the final part of the program where a comparison is made of the two levels of S; if any element of the two differs, an error message is written.

```
C       COMES HERE WHEN N IS NOT A POWER OF 2 TO CHECK SOLUTION           SHRT  30
10      DO 11 J=1,N                                                       SHRT  31
        DO 11 I=1,N                                                       SHRT  32
        IF(S(I,J,1).NE.S(I,J,2)) GO TO 12                                 SHRT  33
11      CONTINUE                                                          SHRT  34
        RETURN                                                            SHRT  35
12      KK=0                                                              SHRT  36
        WRITE(6,13)                                                       SHRT  37
13      FORMAT('0SOLUTION INCORRECT. PAD INPUT DATA WITH DUMMY NODES UNTILSHRT  38
     $ N IS A POWER OF 2')                                                SHRT  39
        RETURN                                                            SHRT  40
        END                                                               SHRT  41
```

SHORT is an example of the advantages of operating on networks in a computer. Problems which are not feasible by hand are quickly and easily solved in a machine with relatively short programs. It should be emphasized, however, that SHORT is a single example of many such programs, and efficient as it is, has disadvantages. For example, the solution matrix contains the shortest distances between nodes, but not the corresponding paths. To determine paths, it is necessary to compare the solution matrix with the original structure matrix. If the network is planar (that is, no crossovers of links occur without a node), then the paths which correspond with distances in the solution matrix are made up of those one-step links which are unchanged in the solution matrix. Determining paths from this information is a trial-and-error procedure by hand, and requires considerable coding and storage space to do in a machine.

A second and more significant disadvantage of SHORT is that it is too slow and requires too much memory for many real-world problems. For example, it would be useful to determine the shortest distance between two points within a matrix comprised of elements with a cost value (such as a highway location problem). In this case, each element is a node, and the space required for S is $2(MN)^2$ where M is the number of rows and N the number of columns in the data matrix. For a matrix 200 by 200, over three billion memory locations would be required, and many times three billion calculations!

There are algorithms which will handle such problems (and provide path information as well) with considerably less storage and time requirements. The interested reader will find several of these in the Collected Algorithms of the Association for Computing Machinery.

Problems

(Suggestions for the Solution of Problems follow.)
1   Draw a flowchart for either the main program for BDRY2 or that for CIRCLE.
2   Rewrite BDRY2 for qualitative data (i.e. BDRY1).
3   Add statements to BDRY2 which compute and print the area of each region.
4   Modify example 13 so that only boundary lines are printed.
5   If you have access to a plotter, make the necessary changes in example 14 so that it will operate with the available system, and modify it so that it plots labels on the outline and each circle, and produces a legend.
6   Rewrite SHORT so that it terminates after there is no further change in the solution matrix.
7   Write a main program which reads a small matrix of values representing a map of costs of crossing a grid cell (such as construction costs or travel time), and converts this matrix to a graph of a network which can be used as a structure matrix for SHORT. Assume all cells except those at the edges are directly connected to their eight neighbours, above and below, each side, and diagonally. (The original matrix should have less than 25 elements.)
8   Take the results of problem 7, and construct a matrix of the same dimensions as the original data, but containing the sum of the shortest distances from it to all other elements of the matrix. The result is a useful accessibility map.
9   Combine INTERP and BDRY2 with a main program which tabulates the number and total value of those elements from INTERP which occur within each region of BDRY2. (This program is useful for such problems as computing the total precipitation which occurs within one or more watersheds using data from a series or irregularly spaced weather stations.)

Suggestions for the Solution of Problems

2   The only change necessary is to specify the vector Z as integer.
3   A vector is initialized to zero at the beginning of the subroutine, and areas determined within the loop in which the map matrix is filled. An additional loop is required to determine the area in the first row.
4   Store the boundary coordinates in an array which may be sent to POINT, example 11.
6   Determine the total of all distances at each step, and compare it with the total at the previous step. When there is no further reduction, control may be returned to the calling program.

7
Some Programming Techniques

An integral part of the programming process for spatial problems is ensuring that the amount of computer time and core memory required is a minimum. Otherwise, such programs are excessively expensive to execute, or are assigned an unnecessarily low priority. In non-spatial applications, savings in time and memory are usually not nearly as important because many fewer data values and computations are involved in a typical problem. As a result, there is sometimes a tendency for individuals with a background in conventional computing to be relatively casual about these problems (although this should not be the case with any competent programmer).

The principal programming techniques for reducing computer time have already been identified and described at several points in this book, particularly those which relate to virtual memory. In this chapter, programming techniques and FORTRAN statements are described which minimize computer memory requirements. In terms of charges and priority, reducing memory space may be just as important as speeding execution time.

The last part of the chapter describes programming techniques which are useful for 'packaging' a program so that it may be easily used by others, and on a variety of data sets in various formats. The conclusion to the chapter and the book is a brief description of four computer languages which are alternatives to FORTRAN.

7.1 Reducing Memory Requirements

A common problem with computer programs which involve spatial data is that they require large amounts of computer memory to store arrays. On smaller machines, this often means that there is simply not enough memory to run the program, and on larger machines, that priority is substantially reduced (and turn-around time increased), or premium charges are assessed. As a result, it is often necessary either to re-formulate the problem, or to attempt to reduce the memory requirements of the computer program.

There are several ways to reduce memory requirements without a corresponding reduction in spatial data stored. Four are described in this section: the COMMON statement, the EQUIVALENCE statement, reduced computer word lengths, and peripheral storage. A fifth technique which can be applied to irregularly shaped study areas is described in the following section.

The COMMON statement is a FORTRAN specification statement which allows two or more program units (main programs, subroutines, or functions) to share a storage area in core memory for variables. It has the form

COMMON name

where name is the variable or variables which are to be stored in the shared area. A COMMON statement is placed with other specification statements at the beginning of each program unit which is to use the storage area, and it may contain dimension information, for example,

```
COMMON ALPHA(25,50)
```

If a variable is dimensioned within a COMMON statement and also appears in another specification statement, then the dimension information may appear in only one of the two specification statements (and it generally should be the first). Thus

```
COMMON JUMP(25),HOLD(10)
INTEGER HOLD
REAL JUMP
```

Variables which are stored in common may not be initialized with a DATA statement or with an explicit type specification, but must be set with a special procedure—the BLOCK DATA subprogram (not described in this book).[1]

The COMMON statement describes how the shared storage area is organized. Variables are stored in the same order in the machine as they appear in the list. This means that COMMON statements in different programs should have variable lists which match in order and type, but not necessarily in names (the same rule as that for argument lists).

COMMON statements may be used to conserve memory when two or more arrays are used in different program units, and can share the same part of a common block without interfering with each other's function in the program. Thus, in example 13, the arrays used in the main program and in BDRY2 to store strings of coordinates and related data can be stored in the same locations as the array used in MAP2 to contain the characters for each line of output. The three COMMON statements required to do this are: (in the main program)

```
COMMON MAP(100,40),NCHARS,NLINES,VERT(2,25),XY(2,100),
$ NP(25),Z(25)
```

(in BDRY2)

```
COMMON MAP(100,40),NROWS,NCOLS,VERT(2,25),RC(2,100),
$ NP(25),Z(25)
```

(in MAP2)

```
COMMON MAP(100,40),NCHARS,NLINES,LINE(131,3)
```

The arrays VERT, XY, NP and Z in the main program and VERT, RC, NP and Z in BDRY2 and the first 300 elements of LINE in MAP2 all share the same part of the common storage area. This is a case where the usual rule that the variable lists must match in type and number of variables is ignored. Note that the array MAP is dimensioned to the same value in all three statements; this is because adjustable dimensions are not allowed for arrays which are in common. Variables in a COMMON statement also may not appear as arguments in a CALL or SUBROUTINE statement, but are transmitted by the fact that they appear in the COMMON statement.

The EQUIVALENCE statement is also a specification statement which causes storage areas to be shared, but refers to variables which are in the same program unit. It has the general form

```
EQUIVALENCE  (a, b, c, ...), (d, e, ...)
```

---

[1] Also not described is a particular type of COMMON statement termed the labelled COMMON.

where *a*, *b*, *c*, *d* and *e* are the names of scalar or subscripted variables (with or without subscripts). For example,

```
DIMENSION A(10),B(20)
EQUIVALENCE(A,B(5))
```

would result in the first element of A and the fifth of B sharing the same memory location. It is important to realize that requiring single elements from different arrays share the same location means that the other elements in each of the arrays are re-aligned, and will probably also share storage locations. In the example above, all ten elements of A will share locations with elements of B.

The principal constraint on the use of EQUIVALENCE statements is that variables in common may not be made equivalent. It is possible, however, to make variables in common equivalent to variables not in common, for example

```
COMMON X(10,10)
DIMENSION Y(10,10)
EQUIVALENCE (X,Y)
```

The EQUIVALENCE statement has several programming applications, but is per-haps most useful for reducing storage requirements in programs where arrays are not used at the same time, and therefore may share the same storage area.

Programming techniques such as COMMON and EQUIVALENCE statements should be the first measures tested for storage reduction because they result in no significant increase in computer time. The next alternative is reducing word length to pack more numbers in a smaller space, with a cost in increased execution time be-cause each number must be converted to standard word size for any calculation.

WATFIV and IBM 360 and 370 FORTRAN provide a capability for reducing word length (for integer variables only) with an extension of the type statement:

INTEGER*2 name

The variables specified in this statement have word lengths of two bytes, half the normal word length of four bytes. The maximum integer value which can be stored in a half-length word is $2^{15} - 1$, or 32 767. This specification doubles the storage area available for spatial data which is within the magnitude limits, the cost being execu-tion time to convert to and from this mode and four-byte integer or real words, and in some cases, the time required and loss of precision involved in scaling real values to fit the magnitude limits. However, unless execution time is exorbitantly expensive, this is recommended as a techniuqe to significantly reduce storage requirements.

In cases where COMMON and EQUIVALENCE statements and half-length words do not reduce a program's requirements sufficiently to fit in computer memory with reasonable charges, it is necessary to store some of the data outside core memory on peripheral storage devices (magnetic discs or tapes) during execution, moving it into the core memory when required for a particular problem. The storage available on peripheral devices depends upon the installation but is normally tens or hundreds as many locations as in core memory. The additional costs are usually charges for trans-mitting data and longer turn-around times. (In some cases, blank tapes or discs must be mounted.)

Peripheral storage is used in FORTRAN with READ and WRITE statements,

statements which control the peripheral device (REWIND and BACKSPACE),[2] and system cards which define the relationship between the program and the peripheral device, in particular the data set reference number. The purpose of the REWIND statement is to position the tape or disc at the beginning of the area reserved by the system card. It has the form

REWIND n

where n is the data set reference number, an integer variable or constant between 1 and 99 (or higher, depending on the installation), but usually excluding 5, 6, and 7. Data are written from core to the peripheral device by what is termed an unformatted WRITE such as

`WRITE(N)X`

where N is the data set reference number and X is the name of an array. No format statement is required with this WRITE because the data are being transmitted to the peripheral device specified by the data set reference number in exactly the same way as they are stored in core memory. Data are moved from peripheral storage to core by an unformatted READ such as

`READ(N)X`

If the array being sent to or read from peripheral storage is less than the size stated in the specification statement, then implied DO's must be used, for example,

`WRITE(N) ((X(I,J),I=1,NROWS),J=1,NCOLS)`

It is important to realize that input and output operations with peripheral storage require that the device be properly positioned. In the sequence of three example statements presented above, the REWIND statement moved the unit to its starting point, and the WRITE placed the array on the tape or disc. The READ then attempts to read values for X, but the tape or disc is now positioned at the end of the array just written. Before the READ can be executed, it is necessary to reposition the device to the beginning of the array. This can be done with a REWIND statement, or with a BACKSPACE. A statement such as

`BACKSPACE N`

results in unit N spacing back one record, where a record is a block of data created by one WRITE statement.

The difference between the REWIND and the BACKSPACE statements is that the former moves the unit all the way back to the beginning of the data set, while the latter moves it back only one record. The BACKSPACE may be put under the control of a DO statement to move back a specified number of records. Moving forward in the data set is also done in terms of records, but can be done only by READ operations.[3] This means that one may have to read a number of records to get to a particular one forward of the present position.

In addition to special FORTRAN programming, peripheral storage usually requires additional system cards which state for a particular data set reference number the type of device to be used, and such information as the area to be reserved and the

---

[2] Sequential data organization is assumed because it applies to both tapes and discs. Direct access storage, which is possibly only on discs, is available on some machines with different and considerably more complicated statements. The interested reader should consult the manufacturer's manuals.
[3] Some compilers provide a SKIP instruction.

disposition of the data set after the program is completed. The nature of these cards is very much dependent on the brand and model of computer, and local ground rules. For purposes of illustration, a typical system card for a scratch area (that is, a storage area in use only for the duration of the program) for large IBM machines is described below.

System cards on large IBM machines are part of the job control language (JCL), a highly cryptic but very powerful language which controls the execution of jobs in user-oriented languages such as FORTRAN. The particular JCL statement required to make available a peripheral storage area is termed a DD or data definition statement. A typical statement is

```
//GO.FT03F001 DD DSN=&SCR,DISP=(NEW,DELETE),UNIT=SYSDA
// SPACE=(CYL,(1,1))
```

Unlike FORTRAN, the blanks in all JCL statements are of critical importance; great care should be taken in punching.

A brief explanation of each of the fields in this DD statement is presented below to give the reader some idea of the kind of information required:

//GO.FT03F001   The // is a signal that this is a JCL statement. The GO signals that any predefined allocation of peripheral storage in the execution step (the GO step) is to be overriden. The number 3 is the data set reference number, and F001 refers to the file number on the peripheral unit (almost always irrelevant for FORTRAN users).

DD   This identifies this JCL statement as a data definition statement.

DSN=&SCR   The DSN (which may also be punched as DSNAME) is a name for this data set. This name is of no real consequence if the data set is to be used only for scratch purposes. If it were to be saved for later use in other programs, the name would be of critical importance for retrieval. The & is a signal that this is a temporary data set.

DISP=(NEW,DELETE)   The DISP field states the disposition of the data set before and after execution. In this case, the data set is specified as being new, and it is to be deleted after execution.

UNIT=SYSDA   This field indicates that the peripheral storage unit to be used is system direct access, a general name for any disc areas which might be available. If a particular type of device is required, it is identified here.

SPACE=(CYL,(1,1))   This specifies the space which is to be allowed for this data set. CYL indicates that the measuring unit is cylinders of a disc. The first 1 indicates that a single cylinder is initially allocated; the second 1 indicates that an additional cylinder is to be allocated if space is exhausted. This is done up to 15 times, thus providing 16 cylinders, far more than would be required for quite large data sets.

In many if not most cases, large spatial data sets are stored permanently on peripheral storage so they can be read into a program without the bother of dealing with thousands of data cards. Creating such data sets should be done carefully to make optimal use of the storage space available and to facilitate rapid transfer into core memory. A data set which has already been created is referred to by a DD statement somewhat different from that used to define a scratch area, for example:

```
//GO.FT03F001 DD DSN=NAME,DISP=(OLD,KEEP),UNIT=2314,
// VCL=SER=X0042
```

In this case no SPACE field is required, because the data set has already been created. The data set name is not lead by an ampersand (&), the DISP field states that this data set has already been created and is to be kept after execution of this program, the UNIT field specifies a particular device (a 2314 disc), and the VOL=SER=X0042 provides the identification of the user's private disc.

It should be clear from this brief description that peripheral storage provides the greatest potential for reducing core storage requirements of the techniques presented. However, it is also the most costly per unit of storage gained because of additional charges for data transmission and use of peripheral machinery, and more complicated programming requirements. COMMON and EQUIVALENCE statements and half-length words provide much less saving in space, but with minimal costs in execution time and additional programming. The following section describes a further technique for reducing memory requirements which applies to a particular kind of spatial data.

7.2   Regularly Spaced Data Values and Irregularly Shaped Study Areas

It has been assumed thus far that any spatial pattern which is described by regularly spaced data values (either because it was originally coded that way, or because it was computed into that form) is best stored and manipulated in the computer as a rect-angular matrix. In many if not most cases, however, study areas in the real world are irregular in shape, often so much so that much of the rectangular matrix is not occu-pied by the study area. This poses no particular programming problems, since cells outside the study area can easily be flagged or coded in some way so that an IF statement will detect them and exclude them from computations, or set them to blank if mapping.

However, a highly irregular study area, or even one which is only slightly irregular but very large, can result in a considerable amount of unused storage space in com-puter memory, and often a substantial waste in computer time due to programs which check each cell before proceeding. In such cases, it is clearly more efficient to use a procedure which stores only those elements within the study area boundary. This is done by placing the map data in a vector, and setting up a table which defines the correspondence between row and column positions in the grid and locations in the storage vector.

If the study area is largely convex in shape, with few or no irregularities along its boundaries (figure 7.1), the storage vector is formed by linking together (concatenating those elements of each column which fall with the region).[4] The vector for figure 7.1 for example, would have as its first two elements, the values at the mesh points located at column one, rows six and seven, then the seven elements from column two, at rows two through eleven, and so on.

---

[4] The vector could also be formed by linking rows, but the result would be highly inefficient on virtual memory machines.

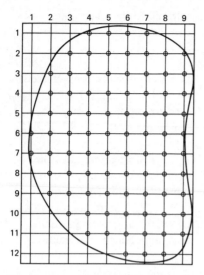

Figure 7.1   An irregularly shaped study area.

The table which defines the correspondence between row and column coordinates and element number in the vector has itself two rows and as many columns as on the original data map (table 7.1). Each column of this table gives for the column with the same number in the map matrix two items of information about the first element of that column which is within the region: the first row in the table is the row number of the element on the original map, and the second row in the table is the number of the element in the storage vector. The relationship between a row and column position i, j on the source map and the location of this element in the storage vector is thus given by

$$\text{map}(i, j) = \text{vector}(i - \text{table}(1, j) + \text{table}(2, j))$$

Study areas with more irregular boundaries or with enclosed areas (figure 7.2) may also be formed into a vector by linking values which occur along columns within the region. However, because each column may cross the boundary several times, the correspondence table described above cannot be applied. Instead, a table must be set up which includes information about gaps in the sequence of elements from each column. This table is also used somewhat differently in that at least part of it must be searched with IF statements whenever a location in the storage vector is required. This

Table 7.1   Correspondence Between Row and Column Coordinates and Element Number in a Storage Vector for figure 7.1

Map Column	1	2	3	4	5	6	6	8	9
Row Number of First Element within Boundary	6	3	2	1	1	1	1	2	2
Number in Vector of First Element Within Boundary	1	3	10	19	30	41	53	65	76

kind of table loop-up consumes far more computer time than using table values directly as subscripts, but is still more efficient than checking each element in a matrix to determine if it is within or without the study area.

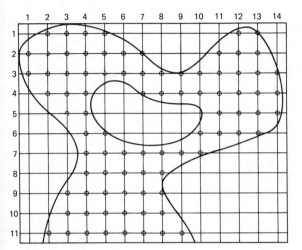

Figure 7.2   A study area with irregular shape and an enclosed region.

7.3   Packaging Programs

Programs which are often used for a particular application are usually designed so that they can be used with a wide variety of data sets in various formats, then are compiled into an object deck by running them with a special option indicated on one of the system cards. This new deck is then used for all subsequent runs of the program, resulting in a considerable saving in computer time because the compilation step is omitted.

The process of designing and writing a program to be flexible in its requirements and to protect the user from error is termed 'packaging' or 'canning' a program. This usually consists of providing for variable size data sets in variable formats, a titling capability, and the ability to select program options. This information is normally described on data cards which precede the main data deck (usually called control cards). A common form for these cards is:

1   parameter or problem card—this contains numbers which define the size of the data set (such as numbers of rows and columns), and the options desired;
2   title card—any information punched on this card is printed as a title on the output;
3   variable format card—the format of the data is punched on this card using standard FORTRAN format codes, enclosed within parentheses.

Let us assume that example 12, INTERP, is to be packaged with SCALE, MAP2 and INTERP as a general computer mapping program for irregularly spaced data points. A possible set of control cards is shown in figure 7.3

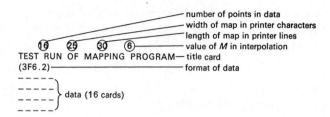

number of points in data
width of map in printer characters
length of map in printer lines
value of *M* in interpolation
title card
format of data

The main program starts with statements which read these cards and check their values:

```
      REAL X(100),Y(100),Z(100),TITLE(18),FMT(18),MAP(50,50)
      READ(5,1)NPTS,NCHARS,NLINES,M
1     FORMAT(4I5)
      IF(NPTS.LE.100.AND.NCHARS.LE.50.AND.NLINES.LE.50
     * .AND.M.GT.0.AND.M.LE.12)  GO TO 2
C WRITE ERROR MESSAGE AND STOP
      WRITE(6,100)
100   FORMAT('1ERROR ON CONTROL CARD. PROGRAM STOPPED.')
      S30P
2     READ(5,3)TITLE
3     FORMAT(18A4)
      READ(5,3)FMT
```

No DO is used when reading either TITLE and FMT because an entire array is being read. (Trailing blanks in FMT have no effect.)

The variable format is used in the subsequent READ by replacing the format statement number with the name of the array which contains the format information. Thus,

```
      READ(5,FMT)  (X(I),Y(I),Z(I),I=1,NPTS)
```

An alternative method for specifying the size of a data set is to use fence cards, as illustrated in examples 13 and 14 and have the program count data values. In most cases, this will not replace the parameter card because it is still necessary to specify information other than the size of data sets. It is a very useful procedure, however, when a program is often re-run with different sizes of data sets.

In some cases, the options desired in a packaged program involve more than simply establishing the values of variables. It is often very useful to be able to change the flow of execution and the subprograms called simply by changing a value on a card. Consider, for example, a more general line printer mapping program than that described above. It will accommodate both irregularly and regularly spaced data values based on either square or rectangular grids, and produces maps with either single characters or blocks of characters representing each grid cell, scaled either by dividing the data range into equal intervals or into intervals which produce an equal number of cells in each class (single character maps only). This program would consist of INTERP, SCALE, SQUEZ2, MAP2, MAP4, MAP2A, SORT and a main program. Table 7.2 lists the options which would be available from such a program, assuming no changes in the subroutines presented in the text. (Several modifications would be required to obtain the full range of possible alternatives.)

Different options in this program require values for different variables. If the data are irregularly spaced, for example, it is necessary to know the number of data values and the value of *M* in the interpolation computation, but this information is of no

Table 7.2 Possible Options for a General Line Printer Mapping Program

Data Spacing	Grid	Map Units (characters)	Scaling Intervals	Option Number
regular	square	single	equal	1
			unequal	2
		block	equal	3
			unequal	
	rectangular	single	equal	4
			unequal	5
		block	equal	
			unequal	
irregular	square	single	equal	6
			unequal	7
		block	equal	8
			unequal	
	rectangular	single	equal	9
			unequal	10
		block	equal	
			unequal	

meaning for regularly spaced data. In many cases, the simplest programming procedure to handle these varying requirements is to require two control cards at the beginning of the data deck, the first describing the major option selected, and the second the values for those variables required for that particular option. Each option would have a separate READ statement for the second of the two initial control cards. This puts extra demands on the user, however, since he must check the program documentation for each option for such details as the format of cards. Where possible, it is better practice to use control cards in a common format.

The information required for the mapping program can be placed on one card with the following format:

columns 1–5: number of the option desired (from table 7.2);
columns 6–10: for options 1–5, number of rows in data deck (assuming it is arranged by rows); for options 6–10, number of lines in output map;
columns 11–15: for options 1–5, number of columns in data deck; for options 6–10, number of characters in each line of output map;
columns 16–20: for options 6–10, number of data points; otherwise blank;

columns 21–25: for options 6–10, the value of M in the interpolation computation; otherwise blank.

A READ statement for this card might be

```
      READ(5,1) JUMP,MN,NM,NPTS,M
1     FORMAT(5I5)
```

It is simpler to omit parameter checking from this initial coding and proceed directly to reading title and format cards as described above. Then an IF statement splits execution into two possible paths, one for irregularly spaced and one for regularly spaced data values. In each case, parameter values are checked as the data are read.

The remainder of the main program consists primarily of IF, GO TO, and CALL statements, and tends to become relatively complicated and difficult to follow. It may be considerably simplified, however, by the computed GO TO statement. It has the form

GO TO $(i, j, k, l), m$

where i, j, k, and l are FORTRAN statement numbers, and m is an integer variable which has a value greater than zero, and less than or equal to the total number of statement numbers within the parentheses. The computed GO TO is an instruction to transfer control to the mth statement number. It has the same effect as a series of IF and GO TO statements. For example

```
   GO TO (16,24,2),K
```

is equivalent to

```
   IF(K.EQ.1) GO TO 16
   IF(K.EQ.2) GO TO 24
   IF(K.EQ.3) GO TO 2
```

The computed GO TO can be very useful where multiple branching occurs in a program.

As a concluding note to this discussion on packaging programs, the reader is cautioned to take special care with matrices which are used in argument lists. Adjustable dimensions cannot be used with these arrays in subprograms unless the number of rows in the data will always be equal to the maximum number of rows in the specification statement in the main program. (The section on adjustable dimensions in chapter 3 provides an explanation for this.) The recommended practice is either to use the standard specification statements, in which dimensions are expressed as constants. In this way, matrix data can occupy any number of rows and columns up to the specified maximum.

7.4   Some Alternative Computer Languages: ALGOL, PL/I, BASIC, APL

The FORTRAN language is entrenched in the computer world. So many people are familiar with it and use it regularly, so many efficient and useful programs are available in FORTRAN, and compilers are available on so many machines, that it is unlikely that an alternative computer language will replace it in the foreseeable future. This is not to say that FORTRAN is the best computer language of those available; in fact, many would argue it is the worst of the major alternatives. Other languages are designed to make them easier to use, more compact, and sometimes more efficient than FORTRAN. Some brief comments on four major languages are presented below.

The decision to formulate the ALGOL language was made in 1958 at a meeting of representatives of the Association for Computing Machinery (an American association) and the European Association for Applied Mathematics and Mechanics. In 1960, a report was issued describing a version of this language now termed ALGOL 60. The purpose of ALGOL is not only to write computer programs to be executed on a machine, but also to represent these programs in a form which may be published and understood with little further explanation. ALGOL is thus intended both for description and implementation of algorithms.

The person familiar with FORTRAN usually has little difficulty with ALGOL because they have a similar range of capabilities and vocabulary, and ALGOL has considerably fewer restrictions. It is not a widespread language, however; only in Europe and the UK has it gained any significant acceptance. In North America it is used primarily by students of computer science.

PL/I (Programming Language One) was developed in the mid-60s with the intent of replacing FORTRAN, COBOL (a business language), and several special purpose languages with a single language which would offer the capabilities of all of these. PL/I is one of the richest and most powerful computer languages ever written, and has been the principal competition for FORTRAN in North America. A feature of particular interest to those concerned with spatial problems is that matrix values in PL/I are stored by rows in the machine, rather than by columns as in FORTRAN. This eliminates the problems with virtual memory discussed in this book. The person familiar with FORTRAN has little difficulty mastering an equivalent subset of PL/I, but has no advantage in learning the additional features.

BASIC and APL are interactive languages. Statements are typed on a remote terminal keyboard and transmitted to the central computer as each is completed. The compiler then checks for syntax errors and immediately sends a message if any are encountered. Program logic may be checked at various points in the development of a program by executing individual statements or groups of statements. Data and programs may be stored on peripheral storage at the central computer to be used when desired from a remote terminal.

BASIC is an acronym for Beginner's All-Purpose Symbolic Instruction Code. It was designed at Dartmouth College for the student or casual user with access to a remote terminal who wishes to learn something about the essentials of programming, and to run small problems. It has fewer capabilities than FORTRAN, but is sufficiently powerful that it is widely used by commercial timesharing firms. Of particular interest for spatial problems is a capability offered by BASIC to do input, output, and assignment statements on matrices without DO statements or their equivalent. Matrix data are stored by rows.

APL (A Programming Language) was invented by Kenneth Iverson in 1962, but was not actually implemented until the late 1960s, and even now (1976) is not always available at large computer centres. It soon will be, however, for APL is one of the most concise and easily learned languages ever devised, yet is more powerful than FORTRAN (and in many ways PL/I). It is particularly well suited for operating on arrays.

References and Selected Reading

Ackoff, R. S., and Sasieni, M. W. (1968), *Fundamentals of Operations Research*. New York: Wiley.

Bernstein, J. (1966). *The Analytical Engine*. New York: Vintage Books.

Berry, B. J. L., and Baker, A. M. (1968), 'Geographic Sampling'. In B. J. L. Berry and D. F. Marble, (1968), *Spatial Analysis*, Englewood Cliffs, N.J.: Prentice-Hall.

Boothroyd, J. (1963), 'Algorithm 201: SHELLSORT'. *Commun. Assoc. Computing Machinery*, **6**, 445.

Haggett, P. (1969), 'On Geographical Research in a Computer Environment'. *Geograph. J.*, **135**, 497–507.

Harbaugh, J. W., and Merriam, D. F. (1968), *Computer Applications in Stratigraphic Analysis*. New York: Wiley.

Hsu, M., and Robinson, A. H. (1970), *The Fidelity of Isopleth Maps*. Minneapolis: The University of Minnesota Press.

Marble, D. F. (1967), *Some Computer Programs for Geographic Research*. Evanston, Illinois: Department of Geography, Northwestern University.

McIntyre, D. B., Pollard, D. D., and Smith, R. (1968), 'Computer Programs for Automatic Contouring'. *Computer Contribution 23*, State Geological Survey, University of Kansas.

Muehrcke, P. (1972), *Thematic Cartography*. Washington, D.C.: Commission on College Geography, Association of American Geographers.

Murray, R. D., editor (1970), *Computer Handling of Geographical Information*. Washington, D.C.: Society of Photographic Scientists and Engineers.

Peucker, T. (1972), *Computer Cartography*. Washington, D.C.: Commission on College Geography, Association of American Geographers.

Robinson, A. H., and Sale, R. D. (1969), *Elements of Cartography*. New York: Wiley.

Shell, D. L. (1959), 'A High-Speed Sorting Procedure'. *Commun. Assoc. for Computing Machinery*, **2**, 30–32.

Shepard, D. (1967), 'A Two-Dimensional Interpolation Function for Irregularly Spaced Data'. *Proc. 23rd Natnl Conf.*, Association for Computing Machinery.

Stuart, F. (1970), *Fortran Programming*. New York: Wiley.

Tarrant, J. R. (1970), *Computers in Geography*. Norwich: School of Environmental Sciences, University of East Anglia.

Tobler, W. R., editor (1970), *Selected Computer Programs*. Ann Arbor: Department of Geography, University of Michigan.

Tomlinson, R. F., editor (1970), *Environmental Information Systems*. Ottawa: International Geographical Union Commission on Geographical Data Sensing and Processing.

Tomlinson, R. F., editor (1972), *Geographical Data Handling*. Ottawa: International Geographical Union Commission on Geographical Data Sensing and Processing.

Wittick, R. (1971), *Geography Program Exchange*. East Lansing: Computer Institute for Social Science Research, Michigan State University.

Appendix

The IBM 029 Card Punch

There are several versions of card punch machines (sometimes referred to as key punch machines). These differ mainly in the character set available, and the manner in which characters are coded into slots on the card. The IBM 029 punch is most commonly used in large computer centres with modern equipment. Some comments on this machine follow which should be adequate for the novice programmer to become operational.

The basic components of the machine are: 1 the keyboard; 2 the card hopper; 3 the punching station; 4 the reading station; 5 the card stacker; and 6 the program control unit.

Keyboard (plate 6)   The keyboard has the standard typewriter arrangement for alphabetic characters except that there are no upper and lower cases for the alphabet; they are all punched as capitals. Depressing the numeric key in the lower left gives the equivalent of typewriter upper case except that on the punch these are numbers and special symbols. Numbers are located on the right side of the keyboard in an adding machine configuration, not on the upper line as in typewriters. (Depressing the alpha key in the lower right is not necessary to return to lower case; in fact, this key will have no effect when depressed unless the program control unit is activated.)

There are several keys such as ERROR RESET, MULT PCH, etc., which can initially be ignored. The keys REL, FEED, and REG are important, however, because they are used to control the flow of cards through the machine. There are also several important switches just above the keyboard which will be discussed later.

Card Hopper   The card hopper is located on the top of the machine at the right. It is loaded by pushing back the sliding plate and placing the cards (up to 500) into the hopper, front printed side forward.

Punching Station   Cards are fed from the hopper down to the punching station, or may be placed in this station manually. The card must not only be located at this station, but registered, that is, held at its left edge.

Reading Station   After punching, the card passes to the left to the reading station. If desired, the card can be ignored and left to pass through (the usual action). However, it is possible to duplicate part or all of this card on to a card at the punching station (if a punching error has been made, for example). This is done by depressing the DUP key for as many columns as desired. You keep track by looking into the window on the program control unit in the upper centre part of the machine. Cards can be loaded manually in this station.

Plate 6   The keyboard of an IBM 029 card punch. (Photograph provided by IBM.)

Card Stacker   The punched cards pass to the left from the reading station and are automatically stacked at the upper left of the machine face down so that they are in proper sequence when removed.

Program Control Unit   Below a hinged panel in the upper centre of the machine is a rotating drum which may be used to automate or program operations on the punch. This is done by punching a card according to certain control conventions, then fastening this card (the 'drum card') around the cylinder. Program control is used if many cards of one format are required (such as a large data set).

Two Important Switches   The main power switch is located below the desk area to the right. The backspace control is located immediately below the reading station. (The backspace control is useful only when nothing has been punched in the previous column on the card; overpunching will have unfortunate consequences.)

Procedure   Turn the machine on. Load a supply of blank cards into the hopper. On the upper part of the panel set the ON/PRINT switch in the up position. Ignore the other switches on this panel except for the ON/AUTO FEED. This switch controls the mode of card flow through the machine. If it is up (ON), whenever REL is pushed on the main keyboard, or whenever column 80 of a card passes the punching station, a new card is fed and registered automatically. To start, push FEED twice to place two cards in the punching station (necessary under the automatic mode).

If the ON/AUTO FEED switch is down, a card is fed by pushing FEED, then registered by pushing REG. After punching, push REL. To pass the card through to the stacker, repeatedly push REL and REG. You will find the switches become inoperative when you try impossible operations. In such cases the ERROR RESET and CLEAR switches may restore things.

Glossary

Algorithm. A set of procedures to be followed in solving any problem of a given type. Computer programs are written to implement algorithms.

Batch processing. A method of operating a computer in which programs are accumulated and run in sequence in one batch. The alternative is shared-time processing.

Choropleth map. A map comprised of regions which are assumed to be homogeneous in terms of some property. If the boundaries of the regions are defined by some property other than that mapped (such as political boundaries for a population density map), the map is said to be a simple choropleth map. If the regions are defined by the property itself (as on a soils map), it is said to be a dasymetric map. Many restrict all these terms to maps based on quantitative data.

Digitizing. The process by which information on maps is converted to numerical form (often by a machine termed a coordinate or graphic digitizer).

Geocoding. The process of coding information so that there is a spatial reference attached (such as a street address or coordinates). Geocoding is a more comprehensive term than digitizing.

Hardcopy. A term used with output devices to describe output which is recorded on paper or film. Softcopy is output displayed on television type devices, but not recorded.

Hardware. The actual machinery which comprises a computer system.

Isarithmic map. A map on which spatial patterns are depicted by isarithms, that is, lines joining points of equal value. A distinction is often made between two types of isarithms: isometric lines and isopleths (Chapter 5). Isarithmic maps must be based on quantitative data.

Job. A program with data and required system cards.

Object program. A computer program which has been compiled from a source program into machine language instructions. The cards of an object program are described as an object deck.

Remote terminal. Machinery for input and output located at some distance from the main computer. A remote terminal can be as small as an attaché case, or as large as a medium size computer.

Shared-time processing. A method of operating a computer in which several programs share the resources of the machine simultaneously.

Software. The operating system, library programs, and other instructions and data sets which are resident on a computer, but not actually part of the machinery (hardware).

Source program. A computer program in a higher level language (such as FORTRAN) before it has been compiled into an object program.

Index

A format, 62
adjustable dimensions, 44–5, 92
ALGOL, 153
AMOD, 83
APL, 154
area calculation, 69
assignment statement, 12–14

BACKSPACE, 11, 145
BASIC, 153–4
binary, 4–5
bit, 4
byte, 4

CALL, 11, 43
carriage control, 17, 33, 51
COMMON, 11, 142–3
comment cards, 20
connectivity matrix, 137
constants, 12–13
CONTINUE, 11, 66
continuation cards, 15, 42
control cards, 19, 149–52
coordinates, 98–9
coordinate transformation, 99–102

DATA, 11, 63
data set reference number, 17
DIMENSION, 11, 26–7
digitizing, 57–61, 66, 98, 106–9, 117–19
distance calculations, 19–22, 52–4, 57
DO, 27–30

E format, 50–1
EBCDIC, 5
EQUIVALENCE, 11, 144
expressions, 13–14

F format, 16
fence cards, 120–1, 132, 150
FLOAT
flowcharts, 35–7
FORMAT, 11, 16–17
FORTRAN functions, 51
FUNCTION, 11, 52

GO TO, 11, 35, 152

hexadecimal, 4–5
Hollerith literals, 63

I format, 16
IF, 11, 34–5
IFIX, 40, 51
initializing, 63-4, 69
isarithms by plotter, 128

JCL, 146–7

loops, 27–30

map accuracy, 61
MOD, 83, 97

overprint characters, 71

page, paging, 29
perspectives, 129–31
PL/I, 153
plotters, 125–8
PRINT, 19
proximal maps, 115

REAL, 53
READ, 11, 15–19, 30–1, 145
RETURN, 11, 43
REWIND, 11, 145

sorting, 88–92
SUBROUTINE, 11, 43
system cards, 19

truncation, 41, 89

variables, 12–13
variable format, 149–50
virtual memory, 29

word (computer), 4–5, 64, 144
WRITE, 11, 17–18, 145